Clashes of Knowledge

Knowledge and Space

Volume 1

Knowledge and Space

The close interrelation of knowledge and power, knowledge and socio-economic development, the conflicts between orthodox and heterodox knowledge systems, and the economisation of knowledge play a decisive role in society and has been studied by various disciplines. The series "Knowledge and Space" is dedicated to topics dealing with the production, application, spatial distribution and diffusion of knowledge. Science Studies, Actor-Network Theory, research on learning organisations, studies on creative milieus, and the Geographies of Knowledge, Education and Science have all highlighted the importance of spatial disparities and of spatial contexts in the creation, legitimisation, diffusion and application of new knowledge. These studies have shown that spatial disparities in knowledge and creativity are not a short-term transitional event, but a fundamental structural element of economy and society.

The volumes in the "Knowledge and Space" series will cover a broad range of topics relevant for all disciplines in the humanities, social sciences and economics focusing on knowledge, intellectual capital or human capital, e.g. clashes of knowledge, milieus of creativity, Geographies of Knowledge and Science, the storing of knowledge and cultural memories, the economization of knowledge, knowledge and power, learning organizations, the ethnic and cultural dimensions of knowledge, knowledge and action, and the spatial mobility of knowledge. These topics are to be analysed and discussed at an interdisciplinary level by scholars from various disciplines, schools of thought and cultures.

Knowledge and Space is the outcome of an agreement concluded by Klaus Tschira Foundation and Springer in 2006.

Peter Meusburger • Michael Welker
Edgar Wunder
Editors

Clashes of Knowledge

Orthodoxies and Heterodoxies in Science and Religion

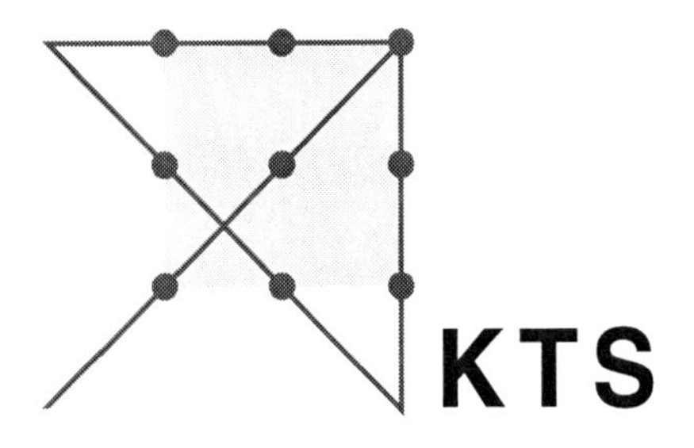

Peter Meusburger
Department of Geography
University of Heidelberg
Germany

Michael Welker
Faculty of Theology
University of Heidelberg
Germany

Edgar Wunder
Department of Geography
University of Heidelberg
Germany

ISBN: 978-1-4020-5554-6 e-ISBN: 978-1-4020-5555-3
DOI: 10.1007/978-1-4020-5555-3

Library of Congress Control Number: 2008921360

Printed on acid-free paper

9 8 7 6 5 4 3 2 1

springer.com

Contents

Contributors

Professor Dr. Günter Abel
Technische Universität Berlin, Institut für Philosophie, Ernst-Reuter-Platz 7, 10587 Berlin, Germany, abel@tu-berlin.de

Professor Dr. Eileen Barker
Department of Sociology, London School of Economics, Houghton Street, London, WC2 2AE, United Kingdom, E.Barker@lse.ac.uk

Professor Dr. Robert Cialdini
Arizona State University, Department of Psychology, Box 871104, Tempe, AZ 85287-1104, USA, Robert.cialdini@asu.edu

Professor Dr. Harry Collins
Cardiff University, School of Social Science, Centre for Study of Knowledge, Expertise and Science, Cardiff, CF10 3WT, United Kingdom, collinshm@Cardiff.ac.uk

Dr. Peter Fischer
Ludwig-Maximilians-Universität München, Institut für Psychologie, Leopoldstr. 13, 80802 München, Germany, pfischer@psy.uni-muenchen.de

Dr. Aileen Fyfe
National University of Ireland, Department of History, University Road, Galway, Ireland, aileen.fyfe@nuigalway.ie

Professor Dr. Thomas Gieryn
Indiana State University, Department of Sociology, 1020 E. Kirkwood Ave., Bloomington, IN 47405-7103, USA, gieryn@indiana.edu

Professor Dr. Wouter Hanegraaff
University of Amsterdam, Faculty of Humanities, Research Group of Hermetic Philosophy, Oude Turfmarkt 147, 1012 GC Amsterdam, The Netherlands, W.J.Hanegraaff@uva.nl

Professor Dr. Peter Meusburger
Universität Heidelberg, Geographisches Institut, Berliner Str. 48, 69120 Heidelberg, Germany, peter.meusburger@geog.uni-heidelberg.de

Professor Dr. Mikael Stenmark
Uppsala universitet, Teologiska institutionen, Box 511, 75120 Uppsala, Sweden, Mikael.Stenmark@teol.uu.se

Professor Dr. Roger W. Stump
State University of New York, Department of Geography & Planning, Albany, NY 12222, USA, rstump@albany.edu

Professor Dr. Dr. Michael Welker
Universität Heidelberg, Wissenschaftlich-Theologisches Seminar, Kisselgasse 1, 69117 Heidelberg, mw@uni-hd.de

Dr. Edgar Wunder
Universität Heidelberg, Geographisches Institut, Berliner Str. 48, 69120 Heidelberg, Germany, edgar.wunder@geog.uni-heidelberg.de

Introduction to the Book Series "Knowledge and Space"

Peter Meusburger

This book is the first in the series entitled "Knowledge and Space," which is dedicated to topics dealing with the generation, diffusion, and application of knowledge. The series stems from the identically titled Klaus Tschira Symposia, a set of ten conferences that began in Heidelberg, Germany, in spring 2006 and that will continue through autumn 2010. These symposia, financed by the Klaus Tschira Foundation, are intended to bring together scientists from various disciplines, schools of thought, styles of reasoning, and scientific cultures in order to bridge some of the gaps between disciplines and to intensify communication beyond disciplinary boundaries. The symposia and the book series focus on the relevance of spatial settings, contexts, and interactions for the generation and diffusion of knowledge; the situatedness of science in space and time; the causes and consequences of spatial disparities of knowledge; the spatial mobility of knowledge; relations between knowledge and power; milieus of creativity; the storing of knowledge and the role of cultural memories; the distribution of knowledge in organizations; the relations between knowledge and competitiveness; the ethnic and cultural dimension of knowledge; the ambivalent relation between knowledge and action; and many other associations between knowledge and space.

These topics play a decisive role in society and are studied in various disciplines and in interdisciplinary research on organizations, creative milieus, learning regions, networks, and clusters. All this inquiry has highlighted the importance of spatiality in the creation, legitimation, diffusion, and application of new knowledge. The widespread assumptions that scientific results can be generated everywhere, that knowledge can be easily and rapidly disseminated throughout the world by electronic communication, and that everybody is able to gain access to the knowledge he or she needs, have proved illusory. In the age of telecommunication, spatial disparities of knowledge have not become irrelevant. Quite the contrary, their significance has increased.

In the second chapter of this volume, it is explained that spatial disparities of knowledge, professional skills, and technology can be traced back to early human history. It is shown that new communication technologies facilitated and accelerated access to freely offered and easily understandable information. They also changed the spatial division of labor, the structure and complexity of organizations, the asymmetry and spatial range of power relations, and the ways in which social

P. Meusburger et al. (eds.), *Clashes of Knowledge*.

systems and networks are coordinated and governed in space. But none of these inventions has ever abolished spatial disparities of knowledge between the centers and peripheries of national or global urban systems.

The generation of various kinds of knowledge (scientific knowledge, orientation knowledge, indigenous knowledge, and other forms of knowledge as described in Chapter 1) was eventually accepted as being situated in time and space. Truth claims, too, came to be seen as being influenced by the social environment. These two changes in thinking sparked new research questions about the meaning of space and place within the processes of knowledge production and dissemination, paving the way for geographies of knowledge, education, and science. Collectively, the contributors to this volume point out that various categories of knowledge are not as mobile in space as is often maintained. The history of science abundantly documents that up to 20 years may lapse before outstanding results, creative ideas, or original theories in one discipline come to be debated or accepted in other disciplines dealing with the same topic or a similar one. Even within a single discipline it may take a decade or more for the gatekeepers of epistemic communities to accept an innovative idea or a revolutionary new theoretical concept. International journals and electronic communication may accelerate knowledge transfer within homogeneous science cultures, within the same discipline, within established networks, or within groups of cooperating disciplines. With few exceptions, however, they seem to do little to accelerate knowledge transfer between disciplines that have no long history of cooperation.

Research on spatial disparities of knowledge and on the relevance of the spatial context for the generation, diffusion, and application of knowledge is an interdisciplinary and even transdisciplinary enterprise. It has become very fashionable in scientific and political debates to demand such a research mode, but it has seldom been adopted in a satisfactory way. The aim of the symposia and of this book series is to offer a platform to those scholars of various disciplines who are aware of these shortcomings and try to go beyond the limits of their own disciplines.

The logo of the Klaus Tschira Foundation (see Fig. 1) serves well as a metaphor expressing our concern about the situation confronting many scholars when they devote themselves to a challenging new research question and find out that problem-solving cannot be confined by disciplinary boundaries. The image presents a solution to the apparently impossible task of connecting nine dots with four strokes from a single marker without losing contact with the writing surface. Any attempt to connect all the dots within the area they define (e.g., within the limits of one's own discipline) is doomed to failure. The only way to solve the problem is to leave the demarcated field by crossing its boundaries three times and approaching the dots from the outside.

The Klaus Tschira Symposia offer an opportunity to cross disciplinary boundaries, and to create new spaces where theoretical concepts, methods, and issues of other disciplines dealing with the generation, diffusion, and application of various forms of knowledge can be intensively disputed. Because creative milieus cannot be planned and governed, such an endeavor is always risky. It remains to be seen whether and under which conditions the spark will jump over the disciplinary gaps,

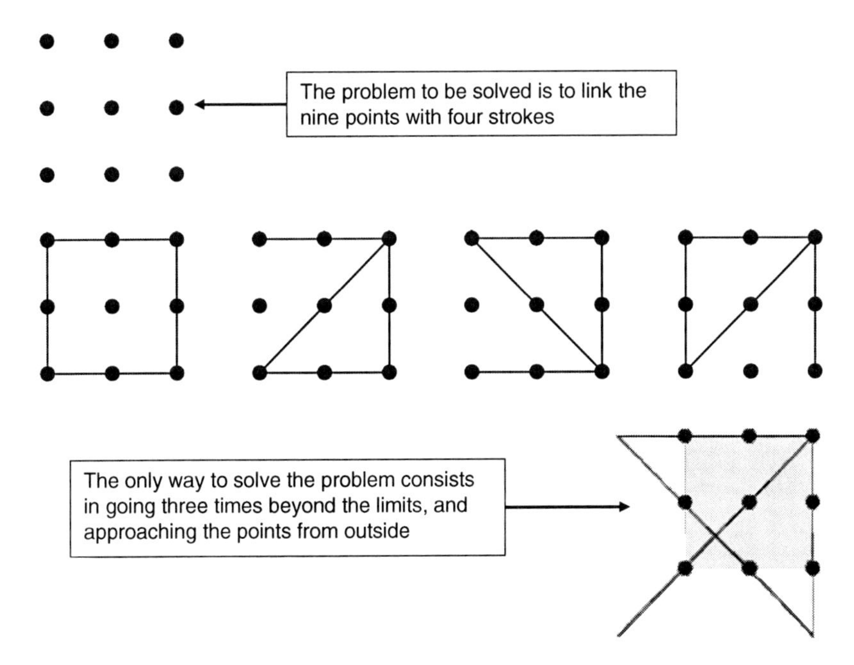

Fig. 1 Connect the nine dots with four strokes

but the experiment is worth a try, and the Villa Bosch offers everything needed for it. We are very grateful to the Klaus Tschira Foundation for providing the "venture capital" for this enterprise. We are equally thankful to Christiane Marxhausen (Department of Geography, Heidelberg University), who is in charge of organizing the first four symposia; to David Antal, who does an excellent job as technical editor of the manuscripts; and to Beate Spiegel, Renate Ries, and Sylke Peters (all of the Klaus Tschira Foundation), who contribute a great deal to the success of the symposia.

Introduction to this Volume

Clashes of Knowledge Inside, Outside, and at the Threshold of Science

Edgar Wunder

The history of science and technology is riddled with examples of outraged ridicule and even outright rejection of new kinds of knowledge and discoveries (e.g., Barber, 1961; Milton, 1996). Highlighting such responses, Truzzi (1990) wrote:

> Some of them are now even silly sounding. Lord Kelvin said that x-rays would prove to be a hoax. Thomas Watson, once chairman of the board of IBM, said in 1943, 'I think there is a world market for about five computers'. ... Ernst Mach said he could not accept the theory of relativity any more than he could accept the existence of atoms and other such dogmas, as he put it. Edison supposedly said that he saw no commercial future for the light bulb. ... Rutherford called atomic power 'moonshine'. (p. 3)

Although the actors in such historical controversies might have perceived the respective disputes as clashes between "knowledge" and "superstition," such a distinction is a quite tricky problem that is not easy to resolve. The notion of "clashes of knowledge," however, would have been regarded as a futile and absurd idea until the beginning of the 19th century. Up to then, knowledge (*episteme*) was generally expected to be certain and infallible, unlike mere opinion (*doxa*). Hence, there could be no "clashes of knowledge" in a self-consistent world. The agent for revealing such infallible knowledge was called *science*.

> Once one accepts, as most thinkers had by the mid-nineteenth century, that science offers no apodictic certainty, that all scientific theories are corrigible and may be subject to serious emendation, then it is no longer viable to attempt to distinguish science from non-science by assimilating that distinction to the difference between knowledge and opinion. Indeed, the unambiguous implication of fallibilism is that there is no difference between knowledge and opinion. (Laudan, 1988, p. 340)

Even worse, the subsequent attempts to compensate for this loss by finding a special epistemological virtue of science—called *the scientific method*—were ultimately unsuccessful. They failed because there was no agreement on what that universal scientific method might be and because all proposals were actually quite disputable descriptions of what most scientists really do (Collins & Pinch, 1998). As stressed by Laudan (1988), "the evident epistemic heterogeneity of the activities and beliefs customarily regarded as scientific should alert us to the probable futility of seeking an epistemic version of a demarcation criterion [of science]" (p. 348). Therefore, "it is probably fair to say that there is no demarcation line between science and non-science, or between science and pseudo-science, which would win assent from a majority of philosophers" (p. 338).

P. Meusburger et al. (eds.), *Clashes of Knowledge*.

Nevertheless, appeals to the myth of the scientific method(s) and the labeling of knowledge claims as "scientific" or "nonscientific" have time and again been powerful rhetorical devices to defend or discredit certain heterodoxies or orthodoxies of knowledge (Bauer, 1992). In scientific communities, as well as in other social contexts, there are always dominant normative systems serving as an instrument to erect the frontier between possibly acceptable knowledge and scientific heresies and to threaten social sanctions against thinkers who dare to cross this borderland (Dolby, 1979). Of course, accepting new kinds of knowledge may necessitate the genuinely unpredictable demolition and reconstruction of whole areas of old knowledge thus far taken for granted, so it is understandable why one is highly motivated to disbelieve unusual knowledge claims. However, few people considering new knowledge claims can afford the time to become familiar with the detailed underpinning argumentation that would make it possible to evaluate their merits properly, so the tendency is to conform to and rely on the norms given in the social environment. Scientists generally do not differ from other people when it comes to being subject to all the biases and self-justifications associated with this herd mentality.

Stigmatization and pejorative labeling reaches its peak when unconventional claims come from outside the established milieu of elite scientists. In the history of science, some powerful gatekeepers have condemned whole areas of research as "pathological science," defined by Langmuir (1968) as "the science of things that are not so," and have exiled their proponents to the remote hell of heretics. Hyman (1980), himself an ardent skeptic to all kinds of unconventional claims, once wrote:

> As a cognitive psychologist, I have tried to reconstruct the thought processes that underlie many of the 'pathological' claims to compare them with those underlying the 'healthy' claims. In most cases I cannot find a difference. And so I was going to argue that there was no 'pathology' in fact involved. The same sort of thought processes that lead some scientists to make claims that Langmuir (1968) calls 'pathological' are just those that have led the very same scientists to make claims, on other occasions, that have found acceptance within the scientific community. ... Langmuir's definition of 'pathological science' as 'the science of things that are not so' is colorful but useless. Much acceptable science falls under this categorization. ... Although Langmuir's definition is not helpful, his cases do stand out as deviant in another sense. They all involve attempts by the scientific community to reject them out of hand—to prevent by any means their entry into the regular channels for scientific evaluation and argumentation. ... [If] there is anything 'pathological' about such cases, the pathology was not to be found by looking into either the truth value of the claims or the manner in which they were justified. Rather the 'pathology' was in the scientific community's reaction to such claims—a reaction that was entirely out of keeping with the scientists' own image of rational, fair, and dispassionate dealing with claims. ... We cannot decide, at least as of now, in advance that a particular claim put forth by a scientist will become one of these cases. This is because my indicants depend upon *how the scientific community perceives and reacts to the claim*. Some claims, even ones that are anomalous and controversial, are accepted as legitimate problems for debate and evaluation within the accepted scientific forum. Others are rejected out of hand. They are not allowed further consideration within the regular forum. It is not the claim as such that I labeled 'pathological', but the manner in which the scientific community responds to and disposes of it. (p. 113)

Such findings challenge the traditional essentialist view, which is still based on hopes for methodological demarcation criteria to reveal a "true" nature of science, to differ science from non-science. But in fact science is, first and foremost, a social institution. To approve such a conclusion, it is not necessary to cling to unlimited relativism or "anything goes" fantasy. It simply has to be acknowledged that what counts as valid scientific knowledge is also always a result of social negotiation and power relations. Conflicts between orthodoxies and heterodoxies in science and other knowledge-generating industries are typically settled, as far as possible, by a spatial separation—by banishment of the dissenters to a foreign social territory and by their exclusion from the resources of one's own networks and institutions (e.g., funding, library use, research and citation cartels, and possibilities to publish work or address conferences). The most important aim of this tendency toward closure is to minimize direct relations between proponents and critics of knowledge claims because even "wrong" knowledge can be infectious.

The social factors in such clashes of knowledge become quite obvious when the status of a scientist in the hierarchy of the scientific community is correlated with his or her readiness to tolerate heterodox knowledge. In an empirical study among 497 members of the American Association for the Advancement of Science (McClenon, 1982; McConnell, 1984), it was found that elite scientists were far more inclined to refuse anomalous experimental results than other scientists or the general population, but only for a priori reasons; familiarity with the relevant research was not an important factor.

Expanding our perspective, it has to be acknowledged that knowledge claims rejected by the scientific community usually also fail to achieve generally accepted legitimacy in modern societies as a whole. Reciprocally, to call something "scientific" is the most popular rhetoric for justifying claims of knowledge. That practice was not always the case and is a result of a long-running expansionist policy of science:

> [The] white patches on the explorers' maps were almost never voids, but territories occupied by other cultures. In the same way, the frontiers of science are not the borderlines between knowledge and ignorance; rather, problems newly taken up by science invariably lead to questions to which other forms of knowledge or belief have already provided answers. (Grabner & Reiter, 1979, p. 67)

Besides clashes of knowledge within science, there are also conflicts at the threshold between science and kinds of knowledge that never have been claimed to be scientific—religion or everyday life experiences, for instance. This kind of boundary work, "a combination of rhetorical and social organizational devices to exclude some people and their knowledge claims from science" (Gieryn, 1983, p. 786), varies contextually and historically (e.g., Livingstone, 1987, 2003). As social scientists, we are unable to understand these clashes of knowledge in an abstract way, ignoring the cultural spaces in which science is embedded.

We cannot even exclude the possibility that the knowledge hegemony science has attained in modern societies toward the end of the 19th century will eventually erode and collapse. There is no "end of history" for either science or democracy. Again and again, competing knowledge systems confront the hegemony of science, and some scholars have already called for intensified efforts in defense of "the scientific worldview" (Perrucci & Trachtman, 1998).

Clashes of knowledge are also abundant between different kinds of knowledge where science, as the hegemonic knowledge system of modern societies, does not seem to be involved at all, as in the sphere of religion (Introvigne, 1995). But the theoretical concepts for studying clashes of knowledge within science can be transferred and applied to other modes of knowledge production as well. The Kuhnian model of paradigm shifts, for example, can be applied to religious change and conversion (Drønen, 2006).

Knowledge created and disseminated by the social institution called "science" (defined here as "scientific knowledge") should not be equated with "analytical knowledge," and "non-scientific knowledge" is by no means the same as "orientation knowledge" (Mittelstrass 1989, p. 21). The function of orientation can be provided by scientific knowledge as well, and even knowledge allocated by non-scientific religious institutions may be of an analytical type. But generally speaking, claims to non-scientific knowledge are more contested than claims to scientific knowledge, and claims to orientation knowledge are more contested than claims to analytical knowledge. One explanation for this tendency is that political, economic, or cultural elites, counterelites, or subcultures can construct and use orientation knowledge systematically to sustain the internal cohesion of their social system and to foster the loyalty of the in-group against an allegedly hostile out-group. This task is facilitated if it can proceed undisturbed by the rather complicated and often normatively cautious considerations of scientists. Consequently, the most severe and violent clashes of knowledge are usually those where non-scientific orientation knowledge is involved. Therefore, contributions to "fundamentalism" and "New Religious Movements" are also included in this volume.

Whether or not knowledge is labeled "scientific," its function of legitimating and fueling processes of social segregation almost always has a spatial dimension. The contributors to this book focus on this spatial dimension and the contextual factors relevant for different kinds of clashes of knowledge. Günter Abel (Chapter 1) and Peter Meusburger (Chapter 2) begin the discussion by trying to clarify some conceptual problems associated with knowledge and space. Thomas Gieryn (Chapter 3), Harry Collins (Chapter 4), and Mikael Stenmark (Chapter 5) then concentrate on clashes of knowledge in the realm of science, with Michael Welker (Chapter 8), Eileen Barker (Chapter 9), and Roger Stump (Chapter 10) focusing on clashes in the field of religion. Aileen Fyfe (Chapter 6) and Wouter Hanegraaff (Chapter 7) discuss clashes of knowledge that jump over the threshold between different kinds of knowledge systems, such as conflicts between science and religion. The last two contributions, by Peter Fischer, Dieter Frey, Claudia Peus, and Andreas Kastenmüller (Chapter 11) and Robert Cialdini (Chapter 12), illuminate clashes of knowledge from a psychological point of view, exploring the question of the circumstances under which individuals may be convinced or manipulated to switch from one knowledge system to another.

References

Barber, B. (1961). Resistance by scientists to scientific discovery. *Science, 134*, 596–602.
Bauer, H. H. (1992). *Scientific literacy and the myth of the scientific method*. Chicago, IL: University of Illinois Press.
Collins, H., & Pinch, T. (1998). *The Golem: What everyone should know about science*. Cambridge, England: Cambridge University Press.
Dolby, R. G. A. (1979). Reflections on deviant science. In R. Wallis (Ed.), *On the margins of science: The social construction of rejected knowledge* (pp. 9–47). Staffordshire, England: University of Keele.
Drønen, T. S. (2006). Scientific revolution and religious conversion: A closer look at Thomas Kuhn's theory of paradigm-shift. *Method and Theory in the Study of Religion, 18*, 232–253.
Gieryn, T. F. (1983). Boundary-work and the demarcation of science from non-science: Strains and interests in professional ideologies of scientists. *American Sociological Review, 48*, 781–795.
Grabner, I., & Reiter, W. (1979). Guardians at the frontiers of science. In H. Nowotny & H. Rose (Eds.), *Counter-movements in the sciences* (pp. 67–104). Dordrecht: Reidel.
Hyman, R. (1980). Pathological science: Towards a proper diagnosis and remedy. *Zetetic Scholar, 6*, 113–121.
Introvigne, M. (1995). The secular anti-cult and the religious counter-cult movement: Strange bedfellows or future enemies? In R. Towler (Ed.), *New religions and the new Europe* (pp. 32–54). Aarhus, Denmark: Aarhus University Press.
Langmuir, I. (1968). *Pathological science*. Schenectady, NY: General Electric Technical Information Series.
Laudan, L. (1988). The demise of the demarcation problem. In M. Ruse (Ed.), *But is it science?* (pp. 337–350). Buffalo, NY: Prometheus.
Livingstone, D. (1987). *Darwin's forgotten defenders: The encounter between evangelical theology and evolutionary thought*. Grand Rapids, MI: Eerdmans.
Livingstone, D. (2003). *Putting science in its place: Geographies of scientific knowledge*. Chicago, IL: University of Chicago Press.
McClenon, J. (1982). A survey of elite scientists: Their attitudes toward ESP and parapsychology. *Journal of Parapsychology, 46*, 127–152.
McConnell, R. A. (1984). *Deviant science: The case of parapsychology*. Philadelphia, PA: University of Pennsylvania Press.
Milton, R. (1996). *Alternative science: Challenging the myths of the scientific establishment*. Vermont: Park Street.
Mittelstrass, J. (1989). *Glanz und Elend der Geisteswissenschaften*. Oldenburger Universitätsreden Nr. 27. Oldenburg, Germany: Bibliotheks- und Informationssystem der Universität Oldenburg.
Perrucci, R., & Trachtman, L. E. (1998). Science under siege: Interest group politics confront scientific knowledge and authority. *Journal of Contemporary Sociology, 35*, 217–242.
Truzzi, M. (1990). Reflections on the reception of unconventional claims in science. *Frontier Perspectives, 1*(2), 3–5.

Chapter 1
Forms of Knowledge: Problems, Projects, Perspectives*

Günter Abel

Types and Forms of Knowledge

Knowledge is a basic word not only in connection with the current discussions of the *knowledge society*. Different forms of knowledge play an important role in people's lives. This is the case with everyday habits, customs, competencies, and practices as well as in science, technology, and institutions of the modern civilized world. Therefore, the different forms of knowledge and in particular their interactions at the interface of human cognition, communication, and cooperation (hereafter, the CCC triangulation) deserve increased attention and should be analyzed and reflected on thoroughly.

The point in this article is not to give an airtight definition of knowledge, as is still the case, for instance, in the endeavor to define knowledge as "justified true belief" (as Plato, 1990c, 201c–201d, did in his *Theaitetos*). Such a definition meets with criticism, as can be made clear by the following two easily construable examples: (a) cases that are not concerned with knowledge but in which the definition given complies with the requirements for knowledge, or (b) cases that deal with knowledge but where the definition does not cover the case. Gettier's (2000) objection to the conception of knowledge as justified true belief is famous. It contains cogent examples of why that definition is incomplete and why it does not represent any sufficient condition for knowledge (see Gettier, 2000).

It is important to see that it is not vital to come up with a subtle revision of the definition mentioned. As soon as the paradigmatic cases have been taken into consideration, it is a question of elucidating different forms of knowledge, which one does not need to define but which one encounters and presupposes by the very act of meaningfully talking, thinking, and acting. The human activities of communicating, thinking, and acting are always already connected with an understanding and a sense of "knowing." In this sense the word knowledge already has its meaning. Meaning does not have to be bestowed on the word by a definition.

---

*The following text is a revised version of Abel (2004, pp. 319–348).

P. Meusburger et al. (eds.), *Clashes of Knowledge*.

But this assumed and implicit meaning of knowing—and of knowledge (including its different sense-critical presuppositions)—has to be made explicit and, if necessary, examined most critically. In the case of scientific knowledge (which is strongly allied to truth and justification), this requisite leads to the claims of knowledge and the critical examination of the requirements for knowledge in the "logical space of reason" (Sellars, 1997, p. 76).

Research into the cognitive and normative roles of knowledge (including the roles of uncertainty and of not-knowing) is relevant not only in its narrow epistemological sense. It is also relevant because it deals with profiles of worlds of knowledge possibly important in the future, with human self-understanding, and with important aspects of orientation in and the future development of modern societies and human forms of life.

Upon closer examination, it is striking how many different meanings the words knowing and knowledge have, meanings that can be found in very different contexts beyond the fields of science and technology. Just think of expressions such as *to be in the know, to let someone know, to know how to help oneself, to the best of one's knowledge, you never know, not to know anything, to know which way the wind blows*, and many more. As always in thinking about knowing, distinctions have to be made. Let us start with three of them.

In view of the variety just mentioned, it is important to distinguish between a narrow and a broad sense of knowing and knowledge. The narrow notion of knowledge refers to knowledge obtained by a methodically well-regulated procedure bound to justification, truth, and verification. It is essential with such knowledge that one be able to talk about it and that it be communicable, transferable, intersubjectively verifiable, and interchangeable *salva veritate*. This notion of knowledge is particularly applicable with reference to the sciences.

The broad notion of knowing and knowledge refers to the ability to adequately grasp what something is about (e.g., what a sentence or a picture is about) on the one hand and the domain of human capacities, competencies, skills, practices, and proficiencies on the other. People are, for instance, very familiar with this domain within their everyday lives (know-how). For the purpose of orientation in the world, we constantly revert to this notion of knowledge and apply it successfully. The use of this broad notion of knowing and knowledge is normally so self-evident that its cognitive, action-stabilizing, and orienting role is not noticed at all until it fails to function smoothly. Such failure occurs when disturbances or problematic situations arise and when it therefore becomes important to reestablish a clear and failure-free situation.

In elucidating the narrow sense of knowing and knowledge, one also has to say a word about other related aspects, for instance, about beliefs, opinions, experiences, skills, verification, justification, and proof. In addition, such elucidations have to include remarks about the possibility and function of error, doubt, not-knowing, and ignorance. Knowing and knowledge are always loaded with preconditions. It is not possible to conceive of knowledge without preconditions, a point already emphasized by Aristotle. There is more to knowledge than we know. For instance, the question of the rationality of forms, practices, and dynamics of knowledge includes more than a relation between theory and observation (which was the dominant aspect within the classical epistemology and philosophy of science), and it includes more than structural characteristics of theories (the latter understood, for instance, as deductive systems of interpretation). Without the broad notion of knowing and knowledge (including the features of un-knowing, not-knowing, not-yet-knowing, and no-longer-knowing), it is not possible to give a comprehensive and satisfying philosophy of human communication, thinking, knowing, perceiving, and acting.

Furthermore, one should distinguish different forms of knowledge. They are very familiar to us because we usually understand the differences that are related to them directly and operate successfully with them. Thus, we distinguish in particular between (a) everyday knowledge (knowing where the letterbox is), (b) theoretical knowledge (knowing that $2+2=4$ or, within classical geometry, knowing that within a triangle the sum of the angles equals 180°), (c) action knowledge (knowing how to open a window), and (d) moral or orientational knowledge (knowing what ought to be done in a given situation).

Across these fields of knowledge (narrow/broad sense; different forms) the following important distinctions and pairs of concepts have to be taken into account: (a) explicit and implicit (tacit) knowledge, (b) verbal and nonverbal knowledge, (c) propositional knowledge (that which can be articulated in a linguistic proposition) and nonpropositional knowledge (that which is not articulable within a *that*-clause), (d) knowledge relating to matters of fact and knowledge based on skills and abilities.

Explicit knowledge is articulated and unfolded, that is, displayable—as in a scientific treatise. In contrast, tacit knowledge means those aspects of knowing that are implicit in situations of perceiving, speaking, thinking, and acting but are not made explicit, are not disclosed at surface. In some sense tacit knowledge does not even have to be made explicit for perception, speech, thoughts, and action to be successful. If one knows that a noise coming from the sky is that of an airplane, one knows a good deal of other things not necessarily explicit in that given knowledge, for instance, that it is possible for machines to leave the earth and that they can move in the air.

Verbal knowledge means knowledge that can be and is articulated by using linguistic expressions. In contrast, the representation of nonverbal knowledge (e.g., pictorial or musical knowledge) is not bound to prerequisites characteristic of verbal forms of knowledge (on pictorial knowledge, see Abel, 2004, pp. 361–369). Forms of nonverbal knowledge are not, for instance, bound to the existence of an alphabet or to a linear arrangement of signs, nor are they bound to the requirement of semantic disjunctiveness of the elements of the system of signs that characterize verbal forms of knowledge.

Propositional knowledge is to be understood as knowledge that can be expressed in a proposition, which, more precisely, can be articulated by means of a *that*-clause (as in *knowing that Picasso was a painter*). In contrast, nonpropositional knowledge is a form of knowledge that cannot be articulated in a *that*-clause. Rather, it is elusive in a characteristic way and cannot really be grasped by words (such as knowing how to understand a bodily movement but not being able to put it into words).

When we speak of knowledge of matters of fact, we mean the form of knowledge that refers to existing objects and events within the world—to tables, cars, molecules, and birthday parties, to that which is the subject matter of a perception, observation, or statement. In contrast, knowledge in the sense of *ability* (know-how) refers to human skills, for instance *knowing how to open a bottle of wine*.

By means of the above-mentioned differences in 1, 2, and 3, a complex matrix and a scaled taxonomy of forms of knowledge can easily be developed. It is a matrix or taxonomy of interest in both a descriptive and a normative sense. Just one of many examples within the field of tacit knowledge is the distinction one can make between the verbal and nonverbal aspects and between those nonverbal aspects that can be propositions and those that cannot, such as the genuine pictorial aspects. With those distinctions one can reconstruct and clarify the correlations between these different forms of knowledge much more precisely, including the possible clashes among them.

Before bringing up some of the problems, projects, and perspectives relating to a comprehensive philosophy of knowledge, I should mention three general aspects that are important when discussing forms of knowledge.

Traditionally, theories of knowledge are understood as answers to the challenge posed by philosophical skepticism. Theories of knowledge and epistemology are—such is the hope—keen to refute the skeptic either through deductive demonstrations (which, for logical reasons, is futile) or through attempts to push the skeptic to the internal limits of reasonable doubt and thus satisfy that person's challenge (which is the much more subtle and successful strategy by far). Conversely, nothing compels the human mind to enter in such a deep sense into the problems of knowledge and epistemology as internal (not external) skepticism does.[1] This statement is true for the skepticism (a) about the outer world, (b) on other minds, and (c) of inner experience, including introspection. When I talk of forms of knowledge in the rest of this chapter, their relation to the problem of philosophical skepticism should *not* be seen at the center of the discussion. The matter is not to refute or to eliminate skepticism by appealing to epistemological certainty. It is rather a matter of critically reconstruing, clarifying, and discussing given forms of knowledge in the sense stated at the beginning of this chapter.

[1] The question of a successful answer to internal skepticism plays a central role in Abel (1995). The answer suggested in that book lies in appealing to the sense-logical presuppositions always accepted in given pragmatic and practical attitudes as well as in the proper functioning of an effective practice of using signs and interpretation. For more details on the antiskeptical capacities of such a philosophy of signs and interpretation (and on its advantages compared to other strategies of refuting skepticism), see Koehne (2000).

The epistemic situation of human beings is not one of an extraterrestrial standpoint or of an absolute conception. It is not a "God's Eye point of view" (see Putnam, 1981, p. 49), from which it would be possible to state in a definitive and generally obligatory way what can be considered metaphysically reliable knowledge and what cannot. As finite beings who are always bound to their particular perspectives within the world, we are cut off from such a standpoint not only for contingent but also for systematic reasons. Such knowledge would not be knowledge of our spirit. Knowledge can only be human knowledge in a human dimension. It cannot be knowledge of a divine dimension.

Explicit attention should be paid to the sense in which the term *form of knowledge*, or rather *forms of knowledge* (guiding this chapter throughout), is to be understood. The suggestion in this chapter is to use form (in line with Kant, 1787/1968, and Wittgenstein, 1980) as a paraphrase for *way* or *mode*. Form*s* of knowledge is then to be understood as ways of knowing/knowledge or modes of knowing/knowledge. Thus, form is *not* to be understood as a ready-made, preexistent, atemporal, and independent system of right order—and that point is crucial. Form is not to be understood as a kind of container into which knowledge has to crystallize to even count as knowledge. Thus form is *not* to be understood as a "universal and atemporal pattern or format of all knowledge."

Nor is it to be understood as a prefabricated or a priori order conceived of as an innate part of knowledge itself, presupposed to exist long before we (as finite and hence perspectivist minds) try to cast such knowledge and *its* "innate and prefabricated form" into one of the forms available to us (e.g., into a language form, a picture form, or an action form).

In both variants of these misleading notions of forms of knowledge (the preexistent atemporal type and the innate type), knowledge is understood as being independent of the form in which it is articulated or manifested. This idea is based on the image that forms of knowledge are just tools, means, instruments, vehicles, vessels, or canals by means of which the contents of knowledge are just transported, communicated, and mediated. But presupposing a pure content of knowledge that is totally unformed is a highly problematic and ultimately inexplicable presupposition. It is at a loss from the very beginning because that which is considered to be the content—the thing to be transported, communicated, and conveyed—cannot be specified without appeal to the underlying system of signs and interpretation. The notion of an epistemological primacy, of a ready-made individuated and specified content of knowledge that is there long before there is any form of signointerpretational articulation, is an empty notion. One should abandon both this notion and the search for a completely unformed content.

But then the interesting question concerning the role and function of forms in knowledge should be asked again in a different way. The thesis is that, for humans as finite and perspectivist beings, contents of knowledge and forms of knowledge *cannot* exist independent of the forms, practices, and dynamics of the underlying representational, interpretational, and sign system. Even for an omniscient and almighty God, forms of knowledge cannot exist completely independent of his signointerpretational practices (for, among other things, such a presupposition would undermine the cognitive almightiness of God).

Forms of knowledge can be regarded as forms (i.e., ways or modes) of articulation and presentation determined by signs and interpretation. They are always based on a history and genealogy of their semantic and pragmatic features. And further changes might take place in the future. This is the case even concerning questions of possible revisions within the field of logic.

Thus, the crucial aspect with regard to the dynamics, justification, and progress of knowledge is not the appeal to something like "The Universal (The One and Only and the Perennial) Form of All Knowledge." What counts much more is whether communication, cooperation, and reference to the world can be continued smoothly, whether actions can follow or not.

The appeal to actions that can connect to and continue communication, cooperation, and reference to the world can also be made fruitful in the realm of questions concerning the generation and the development of knowledge and science. The transition from one epistemological constellation to another—in other words, to the next relevant one—and the dynamics of knowledge included in such a transition cannot be described as though there were a prefabricated rule or set of rules, the core of which one has hit when progress has been made in knowledge and science. If such description were possible, one would just have to figure out this one definite rule or set of rules governing the production and progress of knowledge in philosophy and other sciences. Strictly speaking, it should then be possible to derive and realize the best possible development of knowledge and science from this rule or set of rules. The fact that there is no such access to the optimal development of knowledge and science has been shown by epistemological reflections in contemporary philosophy, as in the thesis of the "underdeterminacy" of scientific theories (Quine, 1969, pp. 302–304), the thesis of the "indeterminacy" of translation of languages in sciences (Quine, 1960, p. 27) and by Putnam's (1983) model theoretical arguments (see also Abel, 1999, pp. 101–120; 2002). In regard to empirical perspectives, an equivalent point is effectively demonstrated by the history of science. There are always different directions of developments possible that can be successfully connected to a given constellation or that can follow it. The development and dynamics of knowledge and of sciences do not work according to principles like The One and Only and External Rule. Rather, they work given the best and creative brains in a particular field at a given time and according to the currently accepted state of the art and its successor states.

Just as the use of forms of knowledge is to be understood in the outlined sense of a possible plurality of ways and modes of knowing/knowledge, there cannot be the one and only linear and a priori history of knowledge and sciences. At the same time, it must also be recognized that the "history of knowledge" and the "philosophy of knowledge," as well as the "history of science" and the "history of philosophy," should no longer be treated independent of each other; they have to go into alliance. In this chapter some problems, projects, and perspectives will be outlined that could be subjects for future research on questions of forms, practices, and dynamics of knowledge.

Information and Knowledge

Information has become a key notion in our times: in the sciences (especially physics, biology, and the cognitive sciences), in the world of the media, and in what is called the new information technologies. As shown elsewhere (Abel, 2004, pp. 290–302), it is also a central notion in philosophy, particularly the philosophy of mind (where the concept of information seems to be able to bridge between cognition and brain, given that information can be realized both physically and phenomenally). Against this background, modern and highly technological societies are often referred to as information societies, and the present age is described as an information age. When information moves into such a fundamental position within these different levels and the aspects mentioned above, it is tempting to grant information priority over knowledge and to grant an information society priority over a knowledge society. At times, the latter is equated with the former. Information is then considered to be knowledge.

If this equation were justified, an information theory of knowledge would be required. One would then expect knowledge to be defined in terms of information. But what has been said elsewhere (see Abel, 2004, pp. 302–304) about the limits of an information theory of the "meaning" of words, sentences, and the human "mind" can also be said about knowledge. In order to focus on the aspects relevant to information, one has to know what one is looking for and what one wants to do with it. Information is always only information in the light of certain knowledge and of a presupposed (syntactic and/or semantic) system of signs and interpretation —not the other way around. From the sense-critical point of view, it is not possible, strictly speaking, to explain what it means to be able to speak of information independent of any form of knowledge, entirely nonepistemically—completely independent, that is, free of signs and free of interpretation. Forms of information are not yet forms of knowledge, and information spaces are not yet knowledge spaces. This point has to be accented despite the fact that in the picture outlined above (which is predominant within the current information- and media-technology society) information is seen to be prior to knowledge, that the possession of information is the possession of knowledge, that forms of information are actual forms of knowledge, and that people initially and primarily live in information worlds.

The following three research desiderata result from this diagnosis: (a) One needs a precise conceptual clarification of the relation between information and knowledge and between information society and knowledge society. Given that both information and knowledge move within signs and interpretations, knowledge now appears as a fourth element beside the clarification of the relations between information, signs, and interpretation (see Abel, 2004, pp. 302–304). (b) The logic and particularly the consequences of the topsy-turvy world outlined above must be analyzed. Although a priority of knowledge over information should be assumed if their relationship is considered systematically, a priority of information over knowledge seems to be prevalent if today's public social opinion is taken as basic. A superabundance of information can perfectly lead to a reduction in knowledge. (c) The specifically normative and the specifically human character of knowledge, which is proper to human beings and which humans strive for by nature according to Aristotle (trans. 1995, vol. 5, book 1, first sentence), must be stressed and spelled out. Because of the dominance of information worlds over knowledge worlds, this excellent virtue threatens to fall by the wayside. In this sense media-mediated information worlds often manifest themselves (particularly in the media) as worlds of opinions and beliefs. So it is also important to spell out the differences between opinions, beliefs, and knowledge, which is also to spell out the prerequisites of knowledge.

Opinion, Belief, Knowledge

Knowledge, as underlined in the first section of this chapter, is a matter loaded with preconditions. This characteristic can be seen in the interrelations between opinions, beliefs, and knowledge. The classical position in this matter is that of Plato, as can be found in his *Theaitetos*: knowledge (*epistéme*) is true belief (*dóxa*) joined with explanation (*lógos*). Within the field of today's epistemic logic, this view is rephrased with the help of the following three elements: A person *S* knows that *p* is the case if, and only if, (a) *S* believes that *p* is the case, (b) *p* is true, and (c) *S* has the justified belief that *p* is the case.

It is true that the connection between knowledge and belief is relevant within the platonic model of knowledge. But the more important point in Plato is that belief and opinion (*pístis*) are to be regarded as mere prephilosophical stages of a truly philosophical and, at best, perfect knowledge (see Plato, 1990a, 454d; Plato, 1990b, 509d–510a, 407b–e). Such a claim, however, does not yet take into account the factual correlation between opinions, beliefs, and knowledge, which plays an important role in theory as well as in actions.

A fundamental breakthrough is found first in the work of Kant (1787/1968). He distinguishes between opinion, belief, and knowledge (the three "modes of holding-for-true" (*Critique of pure reason*, B850)) in relation to the degree of their obligation: (a) Our opinions are not even subjectively obligatory. (b) Our beliefs are a way of holding-for-true, whose obligation is already subjectively sufficient (if one believes in something, one is prepared to accept the consequences). (c) Knowledge is the mode of the holding-for-true, which both subjectively and objectively obligatory.

The crucial point is that the three modes are pyramidal in the sense that they are arranged like a cone that is open at its bottom end (see Abel, 2004, pp. 161–169). The arrangement can be read top-down as well as bottom-up. Top-down means that in order to know something, one must always already have a lot of beliefs and must assume even more opinions. When a person *S* "knows that *p* is the case," then the person also "believes" "that *p* is the case." One cannot conceive that *S* "*knows* that she has a toothache" but does not "*believe* that she has a toothache." Bottom-up means that from the vast realm of opinions one can reach the narrower field of subjectively binding beliefs and from there can arrive at the small terrain of the methodically justified, subjectively, and objectively obligatory binding knowledge.

If one can conceive of (a) knowledge as the third mode of holding-for-true and more precisely as "adequately justified true belief" in the sense mentioned above, and if (b) knowledge is, as emphasized, bound to its articulation and to communicability, and if (c) believing can be construed as a variant of interpreting, then knowledge can be understood as "adequately justified true interpretation based on and determined by a system of signs and interpretation." If the modes of holding-for-true are understood as modes of using and understanding signs and interpretations, then knowledge can be conceived as a specific mode of signointerpretational relations (see Abel, 1995, pp. 317–426; 1999, pp. 304–310).

Knowledge in the narrow sense of the word as well as its epistemic objects are not decreed from an extraterrestrial or God-like point of view. Instead, they are built bottom-up from having opinions to having beliefs and from there up to having knowledge. In this sense one can speak of a genealogy of knowledge growing out of life worlds, a process with an increasing degree of distinctiveness and conceptual normativity. This genealogy is still mirrored even within the epistemic logic, that is, within the logical analysis of the notion of knowledge. In epistemic logic believing is not understood as a momentary mental state or act but as a disposition to act, and knowing (like knowledge) is understood as true belief. We are living in opinion-made worlds, in belief-made worlds and—to a much smaller extent—in knowledge-made worlds.

From this assumption the two following research desiderata result: (a) The reciprocal correlations between opinions, beliefs, knowledge, and, correspondingly, between opinion societies, belief societies, and knowledge societies have to be investigated for their conceptual, notional, and empirical components. (b) Coherent concepts of the objectivity and rationality of knowledge and sciences are required in the light of the above-outlined conditional relations between opinions, beliefs, and knowledge.

Essentialism, Relativism, and Science

In epistemological respects it is crucial with regard to knowledge to escape from the stranglehold of the dichotomy between the claim of absoluteness (essentialism, God's point of view) and the claim of relativism. The strategic task is to get a foothold beyond that dichotomy (as in Abel, 1995). The forms of scientific knowledge, and more precisely the strictness of scientific methods and the validity of their results, are based on the fact that the sciences are tied to the regulative presuppositions of intersubjective communicability, of formal consistency, justification, repeatability, verification, empirical validity, objectivity, and truth.

These presuppositions mean that systems of rules function differently at different levels. First-level and object-oriented rules (e.g., the law of energy conservation in physics) have a role and function different from meta-theoretical and second-level rules (e.g., the quest for simplicity or the regulations for what counts when as a scientific argument), which pertain to methods for revising and, in rare cases, even discarding the first-level rules. And as soon as it becomes relevant to ask what the validity of these meta-rules is based on and how they, in turn, are justified, background worldviews of the sciences and regularities they include come into the picture. The background and the network of those presuppositions and stipulations are characteristically of a public nature (given that they are presuppositions and regularities shared with other speakers and listeners and, in the case of the sciences, with the other members of the scientific community). For that reason, these aspects have been spelled out in more detail elsewhere, within the context of the relations between "Science and the Public" (see Abel, 2004, pp. 391–395; on the relation between rules and meta-rules, see Poser, 2001, pp. 199–207).

The Character of Knowledge with Regard to Worldview, Models, and Symbols

The realm of projects and perspectives that should be subjects of future research (oriented to forms, practices, and the dynamics of knowledge) also includes questions about the role and function of worldviews, models, and language within the sciences. Only a few aspects are hinted at in the following passages (for more details, see Abel, 2004, pp. 117–149 and 370–387).

On the one side, sciences always presuppose a worldview in the way they operate and the way they set up of theories. For example, classical modern physics as shaped by Newton emanates from the background assumption that a physical process is a certain behavior of heavy masses within a coordinate system of absolute time and space. On the other side, this particular scientific worldview shows that sciences not only depend on worldviews but can generate a new worldview (see Mittelstrass, 1989, p. 232). Scientific theory-building and the scientific worldview can be subject to changes and revisions. Thus, to extend the example above, the notion that space and time are absolute coordinates, as thought within the classical physical tradition, is opposed by the view that they can no longer be understood as absolute coordinates and that the space-time is to be seen as a function of the distribution of energy and matter within the universe. Obviously, a fundamental and extremely consequential revision of the underlying worldview is manifested in this contrasting idea.

As to the processes of generating and revising knowledge, it is important to investigate the interaction between, for instance, a scientific theory and its corresponding scientific worldview not only in a narrative and historical but primarily in a systematic way. With regard to the modern scientific establishment of theories, one must focus on the interactions and interpenetrations happening in a revolving-door kind of way between scientific theories, new technologies (e.g., particle accelerators within modern physics or new observation instruments within today's astrophysics, such as the Hubble space telescope), and changing scientific worldviews.

The power of a scientific worldview is also manifested by the profiles of models within science and the way in which they are set up (for additional details on the power of worldviews and pictorial worlds, see Abel, 2004, pp. 117–149). Modeling knowledge plays a key role in articulating, presenting, and storing knowledge. Hence, questions arise concerning the way forms of knowledge are incorporated and articulated by means of models. In this sense models can be understood as knowledge constructions, and, more specifically, as signointerpretational constructions. In other words the term *model* is to be understood in its broad sense as a reconstruction of central characteristics of an object, process, or system. With respect to the triangulation of human cognition, communication, and cooperation (the CCC triangulation) as outlined in the beginning of this chapter, one of the important tasks for future philosophical research lies in elaborating a comprehensive and integrated model theory. Because all setting up of knowledge and of theories is formulated *in* or *by means of* symbols and interpretations, questions of models and of modeling models always presuppose a theory of symbols and interpretation. Hence, it is necessary to broaden the project of an integrated philosophy of models by including a general theory of signs and a general theory of interpretation, both presupposed in modeling knowledge. Finally, a comprehensive model theory would have to be construed as a signointerpretational theory of models. This project is a philosophical desideratum. In this sense, knowledge worlds based on modeling can be viewed as signointerpretational worlds.

Propositional and scientific knowledge is tied to its articulation and presentation within a language. The languages of knowledge and the languages of the sciences are (as stressed above) not just vehicles or containers of pure contents of knowledge. What may count as knowledge at all always depends on the forms and properties of the system of signs and interpretation in use—articulating, formulating, and presenting knowledge. For example, a mathematical formalism describes and articulates the states of a physical system by means of mathematical symbols and parameters, that is, by vectors. Making distinctions beyond this epistemic situation and additionally between the signointerpretational functions on the one side and the states "in themselves" independent of signs and interpretation on the other leads to well-known epistemological problems. At the same time, one begins to recognize the deep sense in which the signointerpretationally determined languages of knowledge are internally intertwined with what counts as the real objects, states, and processes denoted.

The Dynamics of Knowledge

Human knowledge (and correspondingly the realm of not-knowing) and, more specifically, the contents of knowledge change are bound to context, time, and situation. Furthermore, those contents can be expanded, modified, revised, transformed, represented in different ways, arranged in new ones, evaluated differently, characterized by continuities and by discontinuities or ruptures, may depend on the operating scopes of technical instruments, can be forgotten, can completely vanish, but can also be recovered. In short, processes and contents of knowledge are substantially of dynamic nature. Two aspects in particular should be elaborated in more detail: (a) the dynamics of knowledge have (among other things) to be displayed as the dynamics of signointerpretational systems, and (b) the dynamics of knowledge have to be understood and construed in correlation with the other two modes of holding-for-true, that is, in connection with the dynamics of beliefs and of opinions.

With regard to the forms of knowledge, one of the results of my analysis was that the forms of knowledge, that is, the ways and modes of knowledge, are not secondarily and contingently, but primarily and necessarily, dependent on the logical and representational properties of the signointerpretational systems in use and the underlying practices. Knowledge is determined by its signointerpretational character as well as by its time and process character (with its time and process character being possible to rephrase and conceptualize out of the former). Therefore, the question concerning the dynamics of knowing can be treated as a question of the dynamics of the underlying signointerpretational systems.

Within the realm of the narrow sense of knowledge—that is, within, say, the theoretical knowledge and, for instance, the structures of theories—that approach to the dynamics of knowing particularly concerns the sign and symbol relations within the formation of hypotheses and the inductive, deductive, and abductive forms of conclusion. As Peirce (1976, 1977) noted, the latter can be displayed through diagrammatic and pictoriological representation of procedures and notations.

Within the broad sense of knowledge (e.g., tacit and nonpropositional knowledge), this approach especially concerns the relation between an occurring sign (which has become problematic with regard to its semantic and pragmatic features) and continuously comprehensible signs that follow. Given the fact that the relation between a sign and an easily comprehensible subsequent sign is neither logically nor causally deterministic, one is concerned here, too, with the important aspect of creativity in the use of signs, that is, with the new and creative use of signs and interpretations. Thus, one can make a connection between the dynamics of knowledge and creativity (see Abel, 2006). Up to now this connection has been a mysterious, but obviously a constitutive, element for the dynamics of knowledge.

The previously mentioned correlation between opinions, beliefs, and knowledge (see the third section of this chapter) has one important, not yet adequately examined, consequence for the dynamics of knowledge. It is that the dynamics of knowledge are always tied to and involved with their underlying dynamics of believing and with the even broader field of the dynamics of opinions.

Those correlations and dependencies enclose aspects that can be seen top-down as well as bottom-up. Viewed bottom-up, these aspects entail the possibility that changes within the field of epistemic belief (i.e., within the dynamics of believing) may lead to changes in the field of knowledge (i.e., the dynamics of knowledge). Consider the following thought experiment—may heaven or any other powers prevent it from becoming real!—that the modern scientific worldview goes out of style. Further suppose that there are no historians of science left and no testimonials reporting what had happened in the past. Lastly, suppose that an animistic or even a demonic worldview has gradually become accepted anew. It is easy to imagine that completely different scientific contents would then be accepted as contents that count as knowledge. If, as the saying goes, mountains can be moved by faith, then belief can certainly change knowledge.

Seen top-down, these suppositions mean that the dynamics of knowledge may have an influence at the level of the contents of beliefs and can lead to changes there. Revolutionary discoveries in science (such as the heliocentric worldview, evolutionary theory, the theory of relativity, the big-bang theory, today's theory of the human brain, and the genome theory) are obviously examples of the effects in that direction. When fundamental results of scientific research are widely accepted, patterns of beliefs change bit by bit, and eventually patterns of opinions do also. One no longer believes, for instance, that the earth is the center of the whole universe, or that humans have nothing to do with animals in terms of evolution (i.e., that the human genome is totally different from that of animals). Knowledge not only changes the world but can also change beliefs and the realm of opinions.

Of course, the results of empirical verification or falsification of theories, hypotheses, assertions, and models (especially within the empirical sciences) are an important part of the dynamics of knowledge and science. The dynamics of sciences that depend on those factors have been the subject of detailed investigations within recent theory of science (as in discussions about the positions of Karl R. Popper and Thomas S. Kuhn in Lakatos and Musgrave (1970), Laudan (1977), as well as Wolfgang Stegmüller (1979), who calls attention to the dynamics of models (see also Poser, 2001, part B)). If the empirical evidence exceeds a critical limit of previously accepted basic scientific principles, then these principles will have to undergo a revision. This sort of revision has to take place in a way that allows individual occurrences to be reintegrated within the horizon of the changed forms of knowledge and theories. This revision also leads to the fact that hitherto problematic or disparate cases now reasonably fit into the revised patterns of organization. In this sense the dynamics of knowledge are also an interactive balancing and a dynamic reciprocal adjustment of common basic principles and empirical facts. This view can perfectly well be understood in the sense of what Goodman (1983) developed within the field of logic and of what Rawls (1971) called the "reflective equilibrium" (p. 20).[2]

---

[2] Under the heading "equilibrium of understanding" in the philosophy of signs and interpretation, the principle has been applied to the processes of the successful understanding and using of signs. See Abel, 1999, p. 95.

Of course, the aspects concerning the epistemological situation of the sciences (see the fourth section of this chapter) are significantly involved in the dynamics of knowledge and science, particularly in the sense that one cannot assume a stable, rigid, or even fixed relation between (a) first-level object-related methodological fixations (e.g., the axiomatic fixation of the energy conservation law in physics), (b) second-level fixations (by means of which changes of first-level rules can be undertaken and justified, such as the demand of simplicity within the organization of matters-of-fact or the consistency of theories), and (c) third-level regulations (understood as the worldview which governs its time and culture and to which one appeals when justifying the second-level rules as the "ultimate" foundation of the scientist's activities. Those different ways of suppositions and fixations are not bedded on top of or underneath each other in a strict metatheoretical stratigraphic sequence. Instead, they are connected with each other in a revolving kind of way and are like loop-forming processes leading back into one's own beginnings. Those sorts of loop processes are also responsible for the dynamics within the relations between the model character and sign character of knowledge.

It is almost needless to say that the dynamics of knowledge proceed in correlation with time, situation, and context. This relationship is threefold. (a) With regard to form and content, identities and stabilities of knowledge contents tend to cut across time, situations, and contexts. (b) With regard to changes in form and content, ruptures, discontinuities, modifications, revisions, and revolutions are possible and to be noted. (c) Forms and contents of knowledge can or cannot be successfully applied at different times and in different contexts and situations. The clarification of those interrelations has to be seen as a research desideratum, too. Insofar as the point of relevance is the interface of cognition, communication, and cooperation (action)—the CCC triangulation—there is a need to clarify the internal relations between the cognitive, communicative, and cooperative (action-related) dynamics within the different levels of the signointerpretational processes. This clarification then has to be inscribed within the macroperspective of a self-reflection of knowledge and the sciences.

Propositional and Nonpropositional Knowledge

Under the heading "forms of knowledge," the difference and relationship between propositional knowledge (that which can be expressed in a linguistic proposition) and nonpropositional knowledge (that which cannot be articulated in a *that*-clause) are of particular significance. An example of nonpropositional, particularly nonlinguistic, knowledge is visual or pictorial knowledge, that which is incorporated, presented, and expressed in visual experiences and pictures. This form of knowledge is very familiar in human visual experiences, the pictorial presentations and representations that people encounter in daily life, the sciences, the arts, and technology.

Admittedly, it does not seem easy to describe and explain this self-evident familiarity in detail with pictorial elements and structures. Figuratively, one may apply to visual and pictorial experience the point that St. Augustine so aptly made in his well-known answer to the question "But what really is time itself?" As long as nobody explicitly asks what time is—or, similarly, what visual experiences and pictures are—people know very well what they are. But if asked to spell out this self-evident knowledge, one no longer appears to know the answers that used to be a matter of course. In what follows in this section, only a brief remark is made about this point.[3]

Propositional and nonpropositional forms of knowledge are both equally fundamental within the processes of human communication, cognition, and cooperation/action, that is, within the CCC triangulation activities. The classic form of propositional knowledge (both explicit and tacit/implicit) is *knowing* that *p is the case*, in which *p* is an abbreviation for a whole proposition, as in *knowing that Paris is the capital of France*. In contrast, nonpropositional and nonlinguistic knowledge cannot be formulated in predicative terms. This form of knowledge can exist within a subjective or phenomenal state of experience, such as knowing what it feels like to be sad, without the knower yet being able to manage the predicative and terminological use of "sad." The particularity of this form of knowledge is reflected also in the fact that the contrary cannot be the case: From the mere acquaintance with the meaning of the word "sad" it does not follow that one knows what it feels like to be sad.

Forms of nonpropositional knowledge also become manifest in a person's practical skills, pictures, shapes, sounds, gestures, or mental images. This fact is proved by psychological studies on color perception that show how human sensory ability to discriminate and recognize shades of color is far more fine-grained than the human linguistic ability to discriminate colors by means of sentence predicates. In this case the sensoriphemonenal discrimination cannot be reduced to the linguistic and grammatical predication as used in judgments.

An important field of inquiry within future signointerpretational philosophical research will be to describe and elucidate the differences and the interaction between (a) the propositional and the nonpropositional, (b) the verbal and the nonverbal, (c) the explicit and implicit (tacit) forms of knowledge, including the multiple cross-connections of these three pairs of concepts, processes, states of affairs, and phenomena. The clarification of those relations and their internal connections obviously are of fundamental relevance not only within the realm of philosophy but also for all the sciences and arts and for everyday practices, feeling, perceiving, speaking, thinking, and acting. Ultimately, our orientation in the world, to ourselves, and to other persons depends considerably on the successful interplay and interpenetration of those components.

---

[3] Abel (2004, pp. 349–369) deals with the question of whether the signointerpretational approach is capable of accounting adequately for the genuine features of pictures (as opposed to languages, for example); the nonlinguistic character of the pictorial, visual knowledge; and for the internal relation between images and cognition.

Know-How and Rationality

If a person knows how to do certain things, such as how to swim, open a bottle of wine, or hit a volley in tennis—that is, if that individual masters certain abilities, skills, and practices—a question then may be whether he or she does so by referring to or instantiating and executing a "pure form of knowledge." Does the person follow a method or a rule that proceeds in distinct steps, as is the case when following a calculus with preestablished rules? Has the person even found an algorithm (albeit very complex) and then applied it successfully? And does a person who is swimming need to be explicitly conscious of the whole extent and all the facets of what he or she "knows" of swimming (e.g., the individual rules that have to be followed when learning how to swim) in order to be able to swim? Is the person who possesses the know-how of swimming an omniscient superintelligence, someone who makes the decision to either do or leave XYZ after knowing all the relevant cognitive factors with regard to actions and decisions in the sense of maximizing the expectable utility (by using the Bayessian theory of decision)? And do only those decisions and actions that have been accomplished under these circumstances deserve to be called rational? In other words, are only those decisions and actions acknowledged as signs of rationality?

Presumably, it is accepted that the thesis that a person's actual knowledge in the sense of abilities and skills (and the nonlinguistic, nonpropositional, and nonexplicit knowledge manifest therein) *cannot* be adequately described, framed, modeled, and adequately justified by means of the figures mentioned. For example, one does just swim, open the wine bottle, or hit the volley. Were it a conditional requirement for a person to analyze actions and performances in an anticipatory way, that is, were it a condition to separate them into all possible elements and then to assemble and construct those elements in a methodical way as in a calculus in order to start his or her action and performance, that person surely would never start to accomplish acting at all.

Too much explicit knowledge can foil the orienting power of tacit/implicit knowledge and can even lead to disorientation: paralysis by analysis. In many cases, not-knowing (in the sense of not explicitly knowing) can be constitutive for starting as well as for accomplishing an action. Furthermore, satisfactory prognoses of what a person will do next or what that person will leave or do in a similar situation are possible just on the basis of an analysis whose grade of detail does not go beyond what is sufficiently clear with regard to the purpose of the action. If one wants to make explicit as much implicit knowledge as possible before performing a communication, cognition, and cooperation (hence, a CCC activity), the very opposite of successful communication, cognition, and cooperation will often arise. That phenomenon is very familiar. It is also known as the centipede syndrome. As soon as the centipede wants to explicitly show how he is capable of so elegantly coordinating his many legs and move along so smoothly, he gets entangled. To give another example, if the answer to the question of whether or not my tennis partner will hit *this* ball as a volley is made conditionally dependent on the complete analysis of the trajectory of *this* ball at *this* time, including all the other basic conditions with regard to my partner, then neither he nor I will arrive at a conclusion in the face of the never-ending series of ever further fine-graded factors. In other words, while I am still thinking about it, my partner has already scored, or the ball has already hit the ground on his side of the court twice and he has lost the point.

The rationality of the know-how cannot—and that is the important aspect here—be described nor made explicit with regard to a calculus-like or algorithmic and logicomethodically organized sequence of steps (each of which is considered to be definitely determined) and their optimization. Further, it applies to know-how. As Wittgenstein (1980) has emphasized with regard to the actual speaking and understanding of a natural language, it cannot be understood as "operating a calculus according to definite rules" (p. 332, no. 81).[4]

Looking at this scenario from the point of view of a philosophy of signs and interpretation, one hits upon the priority of the performance of signs over the analysis, interpretation, and discursive nature of signs (thus the thesis of this chapter).[5] When our usage of signs in communication, cognition, and cooperation functions smoothly, we follow those signs and rules "blindly" (Wittgenstein, 1980, p. 386, no. 219). That is, those processes cannot be described as though we were following prefabricated criteria and external rules or even laws. We are simply grasping and using the possibilities to continue actions and carry them out smoothly. Whether we succeed or not can simply be seen by whether we are able to proceed without any problems with communication, cognition, and cooperation as well as with their triangulation—for the time being, of course.

In the case of knowing, it is not only with regard to those aspects that the question concerning the relation between knowledge and rationality becomes relevant. It is obvious that rational assumptions and requirements are important for both the broad and narrow notion of knowledge. Speaking of knowledge is internally and sense-logically tied and linked to rational assumptions. With regard to the narrow notion of knowledge, the rational assumption goes along with the characteristics of the notion itself. It is a question of knowledge understood in the sense of methodically obtained conclusions, which are tied to investigation procedures, provability, justifiability, well-grounded reasons, truth, consistency, inferential

[4] Those aspects obviously also refer to questions of the "rationality of decisions." Unlike the classical cognitive studies and the classical economic and rational-choice theories, part of today's cognitive science research refers to "simple heuristics," not to the classical optimizing theorem (see Gigerenzer & Todd, 1999). Conditions of limited time, situation, and knowledge taken into account, "fast and frugal heuristics" can be understood as rules that facilitate rapid decisions, prognoses, and accurate strategies for action, which then can be qualified as rational. Perhaps it might be possible to pull these heuristics out of the actual processes, to model and to teach them, to practice and make them effective for the performances of life with regard to situations under conditions of risk and uncertainty.

[5] For details on this fundamental difference between the performance and the interpretation of signs within the philosophy of signs and interpretation, and on the internal relation of this question to the question of rationality in using signs and symbols, see Abel (1999, pp. 78–100).

certainty, coherence, and empirical validity. As previously underscored in this chapter, rationality assumptions are extremely relevant also with regard to the broad sense of knowledge, that is, to the realm of human capabilities, human competencies, practices, abilities, and skills—in short, to know-how. Admittedly, the important result has been made explicit enough: the rational assumptions within the broad field of knowledge are not just of the same kind and structure characteristic within the narrow field (in the sense of inferential conclusions, conjunctions, and connections that are characterized by explicit logicomethodological rules).

It is important with regard to the broad as well as to the narrow sense of knowledge that neither the rational nor the normative aspect is just of secondary importance but that it is already inherent in people's very speaking, thinking, and acting. If our communication, cognition, and cooperation can continue and proceed without problems, then we obviously have chosen the correct connecting, following, and proceeding action. If not, we find ourselves in problematic situations. We then try to reestablish a state where communication, cognition, and cooperation function failure-free again. In other words, the question of a "correct" use of signs and interpretations has become relevant. Hence, in both cases we are involved in the normativity question right from the beginning. This relation between knowledge and rationality has to be spelled out in a signointerpretational way, for the standards with regard to reestablishment of a failure-free use of signs (for the time being) and to performance of actions cannot be decreed from an external God's point of view. They can only be obtained with regard to those assumptions that we must presume to be satisfied within the failure-free functioning of the communicative, cognitive, and cooperative signointerpretational processes. This dimension of the problem of knowledge is of fundamental importance to our human self-understanding and to our orientation in the world as well as to other persons.

A Unified Theory of Knowledge and Action

Knowledge and action are broader and more fundamental notions than science and theory-building. Neither within the natural and technical sciences nor within cultural, social, media, and cognitive studies is a self-understanding of the sciences able to manage without them. Detailing a unified theory of knowledge and action means placing knowledge and action on common ground. It means neither that knowledge is reduced to action nor that action is reduced to knowledge. One must avoid the praxeological fallacy ("In the final analysis, knowing is nothing but action") as well as cognitivist fallacy ("In the final analysis, action is nothing but determined by knowing").

As mentioned at the beginning of this chapter, one must distinguish between narrow and broad knowledge. But it is also necessary to distinguish between narrow and broad action. Action in the narrow sense can be understood as a conscious, deliberate, goal-oriented, and directed doing. Action in the broad sense can be understood as behavior and response within practical contexts and situations of life.

Considerations have to be based on the reciprocal cross-connection and interplay between knowledge and action within life worlds. Human beings orient themselves within their worlds and with other persons by means of both knowledge and action. And they do that out of and toward the practices of their lives. If life worlds can be characterized as signointerpretational worlds, then one can take the relations included therein as the common and quasi-foundational ground for a unified theory of knowledge and action. More specifically, the desire is for a theory that provides the possibility and the basis for being able to ascribe dispositions of action to a person by means of the interpretation of that person's knowledge. Conversely, the desired theory has to provide the possibility and the basis for being able to ascribe knowledge to a person by means of the analysis of his or her actions and dispositions of actions. Such a theory has been developed within the scope of the general philosophy of signs and interpretation (for its fundamental outlines and details, see Abel, 1999, pp. 299–339). At the level of the formation and elaboration of theories, the theory of knowledge (epistemology) and the theory of action can be formulated as two different, but reciprocally referring, versions within the more general philosophy of signs and interpretation.

Basically, the relation between knowledge and action is a matter of aspects cross-connected in a revolving-door or loop kind of way. Every piece of knowledge has a background in, and is based on, aspects of the practice of life and actions; and if one starts an action, one does so on those assumptions that one considers to be determined and certain, that is, on what one *knows* of the situation in question. More precisely, knowledge can (as noted in the third section of this chapter) be characterized as a mode of holding-for-true and, more specifically, as the third mode of the signointerpretional states of affairs and relations.

The internally interpretative character of knowledge is manifested in other respects as well (in addition to the signointerpretationally determined genealogy of knowledge drawn from the realm of beliefs and opinions), especially in the following five ones: (a) the ascriptions of knowledge; (b) the reports of knowledge; (c) the explanations of knowledge; (d) the methodical organization of knowledge; and (e) the fact that explicit knowledge is (in the above-mentioned sense) tied to its articulation in a symbolic, representational, and inferential system in a deep way that cannot be repealed or jumped over. Ascribing knowledge, reporting about knowledge, giving explanations of knowledge, organizing knowledge methodically, articulating and presenting knowledge—all these activities and processes depend on epistemic perspectives and are performed with reference to determinate contexts and out of points of view and of inquiry. Last but not least, they are occurrences in and by means of signs and interpretations. They can basically be characterized as signointerpretative activities and processes. Knowledge depends constitutively and conditionally (and not only optionally) on a number of signointerpretational aspects.

Given that background, it seems a matter of course to shift from the notiological analysis to the signointerpretatiological analysis of knowledge. Supplying a notiological analysis means stating the truth conditions of sentences like *S knows that p*. This procedure does not get far and, in the twinkling of an eye, it forces one into holistic dimensions. Therefore, it is not implausible to broaden the whole investigation as in the above-mentioned sense and to analyze knowledge, including the following five aspects: (a) the three modes of holding-for-true; (b) the conception of knowledge as "adequately justified true belief" and, more precisely, as "adequately justified true interpretation determined by signs"; (c) the rules of action internally affiliated to the forms of knowledge; (d) the language-impregnated, the symbol-theoretical, and the life-world-determined contexts; and (e) the justification and argumentation with regard to claims of knowledge, a social practice that is shared with other speakers and listeners and is, hence, public in nature.

Actions can be conceived as interpretational constructions as well (as shown by Lenk, 1978). Drawing a line between mere behavioral occurrences (understood as spontaneous movements or as processes of stimulation and reaction) and actions (understood as conscious activities aimed at a purpose) can always be understood as drawing a line that is intrinsically interpretative in character. And the results of such organizational classifications can be labeled "interpretational constructions." By the way, both aspects are already in place when one spatiotemporally localizes and individuates actions and contents of action. And both are signointerpretational processes and results. The sense in which actions can be characterized as perspectival, conjectural, projecting, and constructional—in short, as interpretative—was elaborated by 20 elements of a "signointerpretational theory of action" in Abel (1999). An example is the fact that a person, in taking action, takes up and executes a point of view. Other examples are the facts that scopes of actions are circumscribed and limits are drawn; that selections, preferences, deletions, or completions are made; that newly arising situations are evaluated and put into a given or new taxonomy; and that viewpoints are taken and ascriptions made. The signointerpretational character of these processes is a matter of course for the third-person perspective of an external observer (who ascribes something to someone else) as well as for the first-person perspective of a person taking action. And it is a matter of course not only retrospectively (i.e., not only in reports on or judgments and evaluations of actions) but constituently as well. Actions are performed and executed out of and toward signointerpretationally determined horizons and practices.

Knowledge and action are situated and entrenched within the human practice of life, which is articulated in signs and interpretations. It is important to emphasize the asymmetrical aspect of the fact that a theory is situated within a practice but that the practice is not situated within the theory in the same way. But it is not enough to say that knowledge is entrenched in action. One has to go one crucial step further and see both knowledge and action entrenched within human practices of life, that is, among other things, entrenched within our practices of using and understanding signs and interpretations.

With the help of the heuristic three-level model of the signointerpretational states of affairs and relations, one can adequately take into account and spell out the complex relations between knowledge and action suitably.[6] The distinction between three different levels of the signointerpretational states of affairs and relations can be used to elucidate the specific components, roles, and functions of knowledge and action and in particular to describe their interactions and cross-effects. With regard to these heuristic and methodical instruments, it is possible to elucidate the following four aspects: (a) the entrenchment of knowledge *within* action, (b) the reciprocal cross-relation of knowledge *and* action, (c) the dependency of action *on* horizons of knowledge, (d) the entrenchment of both action and knowledge *in* the signointerpretational practices of human life worlds. Only those four respects and their correlations permit speaking suitably of a unified theory of knowledge and action in a nonreductive way.

The relation between knowledge and action at the primary level of the actual performances of knowing and acting has to be distinguished from the relation between the theory of knowledge and the theory of action. The difference is between first-order and second-order knowledge. One can formulate the latter relation in two ways via signointerpretational relations: first, by concentrating on the theories of knowledge and action with regard to their signointerpretationally determined character (theory internally depends on its articulation and presentation by means signs and interpretations); and second, by focusing on the fact that every second-order kind of knowledge, that is, all knowledge of reflection, depends on the condition that one cannot pursue reflections in a nonsignointerpretational way. If the primary signointerpretational performances (and not the additional interpretations of signs) are seen as the basic processes, then the crucial question with regard to the form of a theory of knowledge and action is how it might be possible to represent this basic performance and process character at the level of articulated reflections and theory-building and how to make it the leading way to form theory.

References

Abel, G. (1995). *Interpretationswelten. Gegenwartsphilosophie jenseits von Essentialismus und Relativismus* [Worlds of interpretation: Contemporary philosophy beyond essentialism and relativism] (2nd ed.). Frankfurt am Main, Germany: Suhrkamp.

---

[6] For more detail, see Abel (1999, pp. 328–336). The model has been developed in Abel (1995) and applied to different fields in *Sprache, Zeichen, Interpretation* (1999) and in *Zeichen der Wirklichkeit* (2004). The model distinguishes different signointerpretational processes, states of affairs, and relations on different levels. Heuristically speaking, there are three levels: (a) the acquisition of something by means of signs and interpretation (such as through a word, a sentence, a hypothesis, or an explanation). From that level we can distinguish (b) habitually entrenched patterns in our perceiving, speaking, thinking, and acting (such as patterns of language, behavior, and customs). From both of these levels, we can distinguish (c) basically categorizing and organizing activities (such as the spatiotemporal localization of events, objects, and persons, and the profile and working of individuation principles within perception, language, and thought).

Abel, G. (1999). *Sprache, Zeichen, Interpretation* [Language, sign, interpretation]. Frankfurt am Main, Germany: Suhrkamp.

Abel, G. (2002). Indeterminacy and interpretation. In D. Føllesdal (Ed.), *The philosophy of Quine*: Vol. 3, *Indeterminacy of translation* (pp. 367–383). New York & London: Garland Publishing.

Abel, G. (2004). *Zeichen der Wirklichkeit* [Signs of reality]. Frankfurt am Main, Germany: Suhrkamp.

Abel, G. (2006). Die Kunst des Neuen. Kreativität als Problem der Philosophie [The art of the new: Creativity as a problem of philosophy.]. In G. Abel (Ed.), *Kreativität* (pp. 1–21). Berlin and Hamburg, Germany: Felix Meiner.

Aristotle (1995). *Metaphysik* [Metaphysics]. In Aristotle, *Philosophische Schriften*. 5 vols. (H. Bonitz, Vol. 5, book 1, Trans.). Hamburg, Germany: Felix Meiner.

Gettier, E. L. (2000). Is justified true belief knowledge? In S. Bernecker and F. Dretske (Eds.), *Knowledge: Readings in contemporary epistemology* (pp. 13–15). Oxford, England: Oxford University Press.

Gigerenzer, G., & Todd, P. M. (1999). *Simple heuristics that make us smart*. Oxford, England: Oxford University Press.

Goodman, N. (1983). *Fact, fiction, and forecast* (4th ed.). Cambridge, MA: Harvard University Press.

Kant, I. (1968). *Kritik der reinen Vernunft* [Critique of pure reason] (reprinted 2nd ed. of 1787). In Königliche Preußische Akademie der Wissenschaften (Ed.), *Kants gesammelte Schriften* (Vol. 3). Berlin: Walter de Gruyter.

Koehne, T. (2000). *Skeptizismus und Epistemologie. Entwicklung und Anwendung der skeptischen Methode in der Philosophie* [Skepticism and epistemology: Development and application of the skeptical method in philosophy.]. Munich, Germany: Fink Verlag.

Lakatos, I., & Musgrave, A. (Eds.) (1970). *Criticism and the growth of knowledge*. Cambridge, England: Cambridge University Press.

Laudan, L. (1977). *Progress and its problems: Towards a theory of scientific growth*. Berkeley, CA: University of California Press.

Lenk, H. (1978). Handlung als Interpretationskonstrukt. Entwurf einer konstituenten- und beschreibungstheoretischen Handlungsphilosophie [Action as an interpretational construct: Outline of a constituent and descriptive theoretical philosophy of action]. In H. Lenk (Ed.), *Handlungstheorien interdisziplinär* (Vol. 2., pp. 279–350). Munich, Germany: Wilhelm Fink.

Mittelstrass, J. (1989). *Der Flug der Eule* [The flight of the owl]. Frankfurt am Main, Germany: Suhrkamp.

Peirce, C. S. (1976). *The new elements of mathematics* (C. Eisele, Ed.), 4. vols. The Hague: Mouton.

Peirce, C. S. (1977). *Semiotic and significs: The correspondence between Charles S. Peirce and Victoria Lady Welby* (C. S. Hardwick, Ed.). Bloomington, IN: University Press.

Plato (1990a). *Gorgias*. In G. Eigler (Ed.), *Plato. Werke in acht Bänden* (Vol. 2). Darmstadt, Germany: Wissenschaftliche Buchgesellschaft.

Plato (1990b). *Politeia* [The Politics]. In G. Eigler (Ed.), *Plato. Werke in acht Bänden* (Vol. 4). Darmstadt, Germany: Wissenschaftliche Buchgesellschaft.

Plato (1990c). *Theaitetos* [Theaetetus]. In G. Eigler (Ed.), *Plato. Werke in acht Bänden* (Vol. 6). Darmstadt, Germany: Wissenschaftliche Buchgesellschaft.

Poser, H. (2001). *Wissenschaftstheorie. Eine philosophische Einführung* [Theory of science: A philosophical introduction]. Stuttgart, Germany: Philipp Reclam jun.

Putnam, H. (1981). *Reason, truth and history*. Cambridge, England: Cambridge University Press.

Putnam, H. (1983). *Models and reality*. In H. Putnam (Ed.), *Philosophical papers*: Vol. 3, *Realism and reason* (pp. 1–25). Cambridge, England: Cambridge University Press.

Quine, W. V. O. (1960). *Word and object*. Cambridge, MA: MIT.

Quine, W. V. O. (1969). Reply to Chomsky. In D. Davidson and J. Hintikka (Eds.), *Words and objections: Essays on the work of W. V. O. Quine* (pp. 302–311). Dordrecht: Reidel.

Rawls, J. (1971). *A theory of justice*. Cambridge, MA: Harvard University Press.
Sellars, W. (1997). *Empiricism and the philosophy of mind* (with an introduction by Richard Rorty and a study by Robert Brandom). Cambridge, MA: Harvard University Press.
Stegmüller, W. (1979). *Rationale Rekonstruktion von Wissenschaft und ihrem Wandel* [Rational resconstruction of science and its change]. Stuttgart, Germany: Philipp Reclam jun.
Wittgenstein, L. (1980). *Philosophische Untersuchungen I* (No. 81). Frankfurt am Main, Germany: Suhrkamp.

Chapter 2
The Nexus of Knowledge and Space[1]

Peter Meusburger

Given the prospects of the Internet and other digital information systems, and the emergence of a borderless world, access to certain forms of knowledge is arguably easier and faster than ever before. Some observers (Cairncross, 1997; Knoke, 1996; Naisbitt, 1995; Negroponte, 1995; Relph, 1976; Toffler, 1980; Webber, 1964, 1973) have gone as far as to predict that advances in communication technology will lead to the death of distance, imperil locational advantages of cities, and make spatial disparities of knowledge irrelevant. Some people assume that scientific results can be generated everywhere, that "objective" scientific results are quickly accepted universally, that knowledge can be easily and rapidly disseminated throughout the world, and that everybody is able to gain access to the knowledge he or she needs. Others argue that knowledge is situated in space and time; that the generation and diffusion of knowledge is affected by the spatial context; that knowledge is built through acts of social interaction; that various types of knowledge spread at different speeds; that knowledge is not only in the heads of individuals but also represented in rules, routines, and architectures of organizations; that knowledge is reified in scientific instruments, machines, and research infrastructure; and that the various carriers of knowledge are never equally distributed in space.

Spatial disparities in knowledge, professional skills, and technology can be traced back to early human history. New communication technologies—from the creation of the first scripts to the invention of paper, the construction of the first printing machine, the innovation of the telephone, and the introduction of digital information systems—facilitated and accelerated access to freely offered and easily understandable information. They also changed the spatial division of labor, the structure and complexity of organizations, the asymmetry and spatial range of power relations, and the ways in which social systems and networks are coordinated and governed in space. But none of these inventions ever abolished spatial disparities pertaining to the production, dissemination, and use of knowledge.

[1] I am very grateful to D. Antal, T. Freytag, H. Jöns, D. N. Livingstone, C. Marxhausen, B. Werlen, and E. Wunder for their inspiring comments on drafts of this chapter and for their challenging questions.

P. Meusburger et al. (eds.), *Clashes of Knowledge*.

Centers of power and knowledge have shifted, but spatial disparities of knowledge have never disappeared. On the contrary, most of these communication techniques enlarged the disparities between the centers and peripheries of national or global urban systems with regard to the distribution of workplaces for highly and marginally skilled persons. The proliferation of printing, the telephone, and electronic communication devices made much of former face-to-face contact dispensable but simultaneously created a demand of new face-to-face contact. Improved communication technology "will lead to more relationships and subsequent face-to-face meetings, as long as some relationships still use face-to-face meetings" (Panayides & Kern, 2005, p. 165; see also Gaspar & Glaeser, 1998).

Many authors have predicted an unproblematic diffusion of codified knowledge through new information technologies or even a notable decrease in spatial disparities of knowledge in the context of globalization and a decline in the importance of proximity (Altvater & Mahnkopf, 1996, p. 269; Henkel & Herkommer, 2004; Machlup, 1962, p. 15; McLuhan, 1964; Radner, 1987, p. 737; Singh, 1994, p. 174; Stehr, 1994a, p. 343; 1994b; Werlen, 1997c, pp. 234, 384; Zare, 1997). However, I argue in this chapter that observers making these attempts to presume or predict the emergence of spatially ubiquitous knowledge make at least one of the following mistakes:

- They overlook the spatial consequences of the vertical division of labor, which become manifest in a spatial bifurcation of skills between centers and peripheries.
- They do not distinguish between knowledge and information and between different categories of knowledge; the distinction between codified and tacit knowledge or between individual and collective knowledge is not sufficient.
- They overlook the importance of the spatial context and spatial interactions in the generation, justification, diffusion, and application of new knowledge.
- They base their empirical evidence about the changing functions of cities on the resident population instead of on the places of work as recommended and demonstrated elsewhere (Meusburger, 1978, 1980, 1996b, 2000, 2001b).
- They disregard the findings of organization theory and underestimate the close affiliation between power and various categories of knowledge. They fail to acknowledge that a spatial system's asymmetry of power relations between center and periphery continually prompts the migration of talent and thus produces, or reproduces, spatial disparities of knowledge.
- They apply a naïve model of linear communication between the sender and receiver of information. When analyzing the process of communicating knowledge from A to B, they overemphasize the producer and codifier of knowledge and neglect the cognitive processes taking place in the receiver of information. They overlook the importance that prior knowledge has for the ability, willingness, or reluctance of potential receivers to accept and integrate certain kinds of information into their knowledge base.

- They focus on codified knowledge as a tradable commodity and fail to notice that the acquisition and application of knowledge is primarily a cognitive process.
- They undervalue the importance of the time dimension in a competitive society. Success in a competitive situation does not depend on knowledge or information per se but on having knowledge before another competitor (agent) does or on receiving information earlier than others.

Some of the standard views that mainstream neoclassical economists had on knowledge were that most of it could be codified and transformed in information, that codified knowledge was a public, tradable, and spatially very mobile commodity, that new communication and transport technologies would diminish spatial disparities of knowledge, that *homo oeconomicus* had access to the knowledge he or she needed for rational decision-making, and that spatial disparities of knowledge were only short-lived. In the last 20 to 30 years, most of these ideas have been largely discredited, not only in science studies, geography of knowledge, and actor-network theory, but also in economics, where they have been gradually replaced by concepts of bounded rationality, evolutionary economics, behavioral economics, learning organizations, new theories of the firm, and the strategic management approach (for an overview see Amin & Cohendet, 2004; Gigerenzer, 2001; Gigerenzer & Selten, 2001; Simon, 1956). The classical thinkers in sociology, too, once believed that scientific truths are generated independent of any local context. Durkheim (1899/1972) distinguished religion from science precisely in terms of the situatedness of the former and the placelessness of the latter (Gieryn, 2002c, p. 45).

In science studies, in the geographies of knowledge, science, and education, and, recently, in economics, scholars argue that new knowledge is created in particular places and contexts, often through interaction with other places and through relations within space. They do not regard spatial disparities of knowledge as short-term transitional phenomena. On the contrary, spatial disparities of knowledge are understood as a fundamental structural phenomenon of any society with a highly developed division of labor. In a dynamic and competitive society, the search for and acquisition of knowledge and skills are continuous processes that never finish. In many situations, it is not knowledge per se that counts but rather the possession of prior, specialized, unique, superior, or rare knowledge. It is a head start in generating and applying new knowledge that counts. Mainly for that reason, some kinds of knowledge are kept secret as long as possible or necessary (Brunés, 1967; Konrád & Szelényi, 1978). The fact that a considerable amount of knowledge is kept secret for a certain span of time has aroused much less interest in geography and economics than has the knowledge exchange in and between firms.

All new knowledge starts as local knowledge. Locally produced knowledge as competence of locally situated actors becomes widely disseminated knowledge only if it is shared with others, recognized by epistemic authorities of the relevant domain, and proved useful. If a scientific experiment is only successful at one place and cannot be replicated elsewhere, it gains no credibility (Collins, 1983, 1985; Gieryn,

1999; Livingstone, 2002, 2003; Shapin, 2001). A spatial context not only influences the generation of knowledge, it also strongly affects the justification, legitimation, dissemination, acceptance, interpretation, and application of knowledge. Science and the humanities are replete with examples illustrating the extremely long time it took for highly creative ideas, new research questions, methods, and theoretical concepts to be perceived and accepted by the epistemological centers of the relevant disciplines. It took 11 years until Max Planck's quantum theory was accepted by the leading physicists (Polanyi, 1985, p. 63). A spatial diffusion of knowledge does not guarantee that readers will interpret that knowledge as intended by the writer. Darwin's theory of evolution, for instance, was interpreted very differently, depending on the country in question (see Livingstone, 1987, 2003; Numbers & Stenhouse, 2001; Stenhouse, 2001). Alexander von Humboldt's work, too, was variously received from one land and period to the next (Rupke, 2005).

Some kinds of knowledge diffuse very slowly in space and arrive only at relatively few places. Among these forms are implicit knowledge, nonverbal knowledge (e.g., the competence to play piano), nonpropositional knowledge (a type of knowledge that cannot be articulated in a that-proposition, such as knowing how to understand a bodily movement; see Abel, p.14), and embedded or encultured knowledge (Blackler, 2002) arising from socialization and acculturation in specific cultural settings or shaped by stable relationships in organizational routines and interpersonal relationships. Some contents of knowledge (e.g., gene technology, nuclear energy, and interpretation of certain "historical facts") are opposed by political elites and therefore do not circulate in certain areas. In other words, the generation and diffusion of knowledge is affected by many influencing factors, and any delay or impediment in the diffusion, acceptance, and application of knowledge produces new spatial disparities of knowledge, at least temporarily.

This chapter is an examination of various relations between knowledge and space and debates some of the reasons why spatial disparities of knowledge evolve and why they are so persistent. Before discussing relations between knowledge and space and explaining some of the reasons underlying spatial disparities of knowledge, I inquire into concepts of space, place, spatiality, and spatial scales. The proper consideration of spatial concepts and space-time has crucial effects upon the way theories and understandings are articulated and developed (see Harvey, 2005, p. 100; Kröcher, 2007) and the way the nexus between knowledge and space can be explicated. I also review the significance of spatial contexts for generating, legitimizing, controlling, manipulating, and applying knowledge, especially scientific knowledge, and propose a model for the spatial diffusion of various types of knowledge. The chapter presents a brief report on the developmental paths and research interests of the geographies of science, knowledge, and education and discusses some of the key questions that are decisive for building bridges between the discourses of various disciplines.

The Significance of Spatial Patterns, Spatiality, and Spatial Contexts in Social and Behavioral Sciences

Conceptions of Space and Place

Until the 1970s many scholars of human geography and other social sciences took space for granted. Its existence was so obvious that it was not a matter of heated theoretical debates. Early concepts of space resembled more or less the notion of a confined container enclosing physical-material objects, human beings with ideas and attributes, animals, and artifacts. Searching for spatial laws, spatial factors, and purely spatial processes, devotees of quantitative geography and regional science defined their discipline as the science of the spatial and argued that the explanation of geographical patterns lay within the spatial dimension. Social aspects were widely neglected (for a critique of this position see Massey, 1985, p. 11). A severe blow to this traditional concept fell at the end of the 1960s, when some of the leading quantitative geographers declared that spatial patterns were overdetermined when it comes to the problem of inference, or the *explanation* of the manner in which spatial structures were created (see Barnes, 2004, pp. 589–590). Harvey (1969), once one of the outstanding quantitative geographers, made a radical shift away from positivism and proclaimed in 1972:

> [Geography's] quantitative revolution has run its course. [It tells us] less and less about anything of great relevance ... There is a clear disparity between the sophisticated theoretical and methodological framework which we are using and our ability to say anything really meaningful about events as they unfold around us ... In short, our paradigm is not coping well. (p. 6; see also Barnes, 2004)

Other critics of traditional concepts of space as a "taken-for-granted world" (Ley, 1977) drew on phenomenology and action theory. Ley claimed that geographers should not be interested only in spatial patterns of social facts and processes or in the subjective perception of places but also in the subjective constitution of the meaning of "place" (see also Werlen, 1997a, p. 647). On the basis of Heidegger's existential phenomenology, Pickles (1985) elaborated a perspective in which not "space" but rather the appropriate interpretation of human spatiality should be the aim of social geography (see Werlen, 1997a, p. 648; 1999, p. 5).

Werlen (1993, 1999), a proponent of subject-centered action theory, argued that space does not exist as a material object, or as a consistent object of empirical research. "It is ... rather a formal and classificatory concept, a frame of reference for the physical components of actions and a grammalogue for problems and possibilities related to the performance of action in the physical world" (Werlen, 1993, p. 3). For him "materiality only becomes meaningful in the performance of actions with certain intentions, and under certain social and subjective conditions" (Werlen, 1993, p. 4). He insists that scientific investigation had to center on subjects, not primarily spaces, regions or spatiality. He starts from a perspective "that emphasizes subjective agency as the only source of action and hence of change, at the same time as it stresses that the social world shapes the social actions that produce it" (Werlen, 1993, p. 3). Werlen calls for a rigorous categorical shift from "space" to "action" or from "a geography of the things" to "geographies of the subjects" and to "everyday regionalizations" (Werlen, 1993, pp. 2–4, passim). He argues that the relational concept of place is about human agency and the interplay between structure and agency (Werlen, 1993, p. 253; 1995, 1997b, c, 2004a, b).

Lefebvre (1991) sees space from the opposite perspective. For him the social relations of production have a social existence only insofar as they exist spatially; they project themselves into a space while producing it. In other words, all social relations are spatial, and all spatial relations are social (Markus, 2006, p. 321). It has recently become very fashionable in postmodern geography to relate the reassertion of space in social theory (Soja, 1980) to Lefebvre. However, Schmid (2005, p. 13) argues that the reception of Lefebvre in most cases is very superficial and full of misinterpretations.

Representatives of material semiotics have tried to "bring materiality back in and to see places as generated by the placing, arranging, and naming the spatial ordering of materials and the system of difference that they perform" (Hetherington, 1997, p. 184). In the course of the spatial turn (the discovery, or rediscovery, of the importance of space and spatiality) in the social sciences and humanities, the discussion about correct concepts of space has become even more controversial (see Kröcher, 2007; Lippuner & Lossau, 2005; Löw, 2001; Meusburger, 1999; Schlögel, 2003; Werlen, 2007).

According to Hayden (2001, p. 11451) place is "one of the trickiest words in the English language." It carries the resonance of location as well as of a position in a social hierarchy. A "sense of place" can be an aesthetic concept or can settle for local distinctiveness. A phrase like "knowing one's place" can imply power relationships or a sense of belonging or an emotional attachment to a place. Social relationships and memory are intertwined with spatial perception, with sites of memory, landmarks of triumphs and defeats, massacres, or civil rights. However, the human attachment to places is so complex that it defies simplistic explanations (see Hayden, 2001, pp. 11451–11453). One of the functions of place is gathering and holding together things, experiences, histories, and thoughts, enabling copresence and triggering or releasing memories. According to Casey (1996),

> a given place takes on the qualities of its occupants, reflecting these qualities in its own constitution and description and expressing them in its occurrence as an event: places not only are, they happen. … Places are qualified by their own contents, and qualified as well by the various ways these contents are articulated (denoted, described, discussed, narrated, and so forth) in a given culture. (pp. 27–28)

Harvey (1973, 2005) tries to bridge some of the gaps between different concepts of space. He first distinguishes between three types of space: absolute, relative, and relational.

> If we regard space as absolute it becomes a "thing in itself" with an existence independent of matter. It then possesses a structure which we can use to pigeon-hole or individuate phenomena. The view of relative space proposes that it be understood as a relationship between objects which exists only because objects exist and relate to each other. There is

> another sense in which space can be viewed as relative and I choose to call this relational space—space regarded in the manner of Leibniz, as being contained in objects in the sense that an object can be said to exist only insofar as it contains and represents within itself relationships to other objects. [Absolute space] is fixed and we record or plan events within its frame ... [I]t is usually represented as a pre-existing and immovable grid amenable to standardized measurement and open to calculation. ... Socially, it is the space of private property and other bounded territorial designations (such as states, administrative units, city plans and urban grids). (Harvey, 2005, p. 94)

The relative notion of space rests on Einstein's argument that all forms of measurement depended on the frame of reference of the observer (Harvey, 2005, p. 95). As for the relational view of space,

> there is no such thing as space or time outside of the processes that define them. ... Processes do not occur *in* space but define their own spatial frame. The concept of space is embedded in or internal to process. This very formulation implies that, as in the case of relative space, it is impossible to disentangle space from time. We must therefore focus on the relationality of space-time rather than of space in isolation. The relational notion of space-time implies the idea of internal relations; external influences get internalized in specific processes or things through time ... An event or a thing at a point in space cannot be understood by appeal to what exists only at that point. It depends on everything else going on around it. (Harvey, 2005, p. 96)

Similar arguments about the fluidity of space or the daily making of space are put forward by other authors such as Massey (1999a), Werlen (1987, 1993, 1995, 1997b), and Löw (2001). "We are constantly making and re-making the time-spaces through which we live our lives" (Massey, 1999a, pp. 22–23). Thrift (1999) summarized this issue with the following words: "Like societies, places can be made durable, but they cannot last" (p. 317). Some authors argue that it is the relational ordering of living entities and social goods, the connections between them, and the symbolic meaning of them that constitute space (Löw, 2001).

For Harvey (1973),

> space is neither absolute, relative or relational in itself, but it can become one or all simultaneously depending on the circumstances. The problem of the proper conceptualization of space is resolved through human practice with respect to it ... The question 'what is space?' is therefore replaced by the question 'how is it that different human practices create and make use of different conceptualizations of space'. (p. 13)

As human beings, we are inescapably situated in all three frameworks of space simultaneously. The three concepts are in dialectical tension with each other. "Ground Zero" is an absolute space at the same time as it is relative and relational (see Harvey, 2005, pp. 98–99).

Inspired by Cassirer (1944), who distinguished between organic, perceptual, and symbolic space, and by Lefebvre (1991), Harvey (2005, pp. 101–102) works with a second categorization of space, differentiating material space (experienced space), representations of space (conceptualized space, space as conceived and represented), and spaces of representation (the lived space of sensations, the imagination, emotions, and meanings incorporated into the way people live day by day). Material space is the space of perception open to experience, physical touch and sensation. It is the world of tactile and sensual interaction with matter, the space of experience. The abstract representation of material realities is achieved through maps, pictures, graphs, words, and other means of communication:

> The physical and material experience of spatial and temporal ordering is mediated to some degree by the way space and time are represented … The spaces and times of representation that envelop and surround us as we go about our daily lives likewise affect both our direct experiences and the way we interpret and understand representations. (Harvey, 2005, p. 102)

Combining these two categorizations of space, Harvey (2005, pp. 105, 111) draws a matrix within which points of intersection suggest different modalities of understanding the meanings of space and space-time. Although it is the dialectical relation between the categories that really counts for Harvey, each of the nine categories of space can become relevant (admittedly to a different degree) when studying the nexus between knowledge and space. Werlen (1993, 2007) certainly does not agree with Harvey's conceptualization of material space. In his view, Lefebvre's formulation "involves a double reification: the reification of space, and the reification of relations of production" (Werlen, 1993, p. 4). He opposes the assertion "that space or materiality already have a meaning in themselves, a meaning that is constitutive of social facts. Materiality only becomes meaningful in the performance of actions with certain intentions, and under certain social (and subjective) conditions" (p. 4).

It is not my intention to summarize the extensive and controversial academic debate on concepts of place, space and spatiality or structure and social action in this chapter (see Barnes, 2004; Gieryn, 2000, 2002c; Gregory, 1994; Günzel, 2006; Hard, 1999, 2002; Harvey, 1972, 1973, 2005; Hasse, 1998; Hayden, 1995, 2001; Jahnke, 2004; Klüter, 1986, 1999, 2003; Koch, 2003; Kröcher, 2007; Lefebvre, 1991; Lippuner, 2005; Lippuner & Lossau, 2005; Löw, 2001; Massey, 1999a, b, 2005; Meusburger, 1999; Pred, 1984; Relph, 1976; Sack, 1980; Schatzki, 2007; Schmid, 2005; Soja, 1980, 1985, 2003; Thrift, 1983, 1985, 1999; Tuan, 1977; Weichhart, 1996; 1999, 2003; Werlen, 1987, 1993, 1995, 1997b, c, 2004a, b, 2007). Nor is it possible to condense the debates about the constitution of "reality," on the relation between the "social" and the "material" or between structure and agency in a short paragraph.

Most authors will probably agree with one of the following definitions:

1. Space is the result (product) of social relations (Harvey, 1973; Werlen, 1987, 1993).
2. Space is the relational ordering of social goods and people (Löw, 2001).
3. Space is a means of perception, a performative act (Löw, 2001). Space is an element of social communication (Hard, 2002; Lippuner, 2005, p. 129; Werlen, 2004b).
4. Space is a semantic concept of order in which the physical-material space serves as an element of order and bears a semantic meaning (Miggelbrink, 2002, p. 344).

To me, it makes little sense to maintain that one concept (or understanding) of space or place is in principle more relevant or adequate than another. Those relative qualities of the concepts depend on the research topic, the scale of investigation, and the nature of the phenomena under study. The meaning of place, the link between place and function—that is, between the sign and its object (Pucci, 2006, p. 169)—and the way people interpret space and orient themselves in it (Wassmann, 1998, 2003) vary across culture and time periods.

Taking into account the results of psychological experiments on unconscious perception, implicit learning, implicit memory, and automatic (uncontrolled) reactions (Merikle & Daneman, 1996, 1998; Reber, 1993), one asks whether geography of knowledge, human ecology, or action theory can ignore unconscious cognitive processes any longer. In my view, future research on the relations between environment (spatial context) and actions should also include the role of subliminal or unconscious perception, implicit learning, implicit memory, and procedural knowledge (Anderson, 1983; Merikle, 2000; Merikle & Daneman, 1996, 1998; Reber, 1993). According to Merikle (2000), "subliminal perception occurs whenever stimuli presented below the threshold or limen for awareness are found to influence thoughts, feelings, or actions. ... [T]he term has been applied more generally to describe any situation in which unnoticed stimuli are perceived" (p. 497). Psychological experiments during anesthesia have shown that unconsciously perceived information can remain in the memory for a considerable time. This work suggests "that unconscious perception may have relatively long lasting impact if the perceived information is personally relevant and meaningful" (Merikle & Daneman, 1998, p. 16). According to Reber (1993), implicit learning is "the acquisition of knowledge that takes place largely independently of conscious attempts to learn and largely in the absence of explicit knowledge about what was acquired" (p. 5). Consciousness and phenomenological awareness are late arrivals on the evolutionary scene (pp. 7, 86). "Hence, consciousness and conscious control over action must have been 'built upon' ... deeper and more primitive processes and structures that functioned, independently of awareness" (p. 7). According to Reber (1993) and Merikle and Daneman (1998), many psychological experiments on implicit learning have shown that people acquire complex knowledge about the world independently of conscious attempts to do so. Unconscious cognitive processes apparently tend to be more robust and basic than explicit cognitive processes (Reber, 1993, p. 18).

The findings on implicit learning are paralleled by those on implicit memory. Drawing on these experiments, Anderson (1983) distinguishes between declarative knowledge, which is knowledge that people are aware of and can articulate, and procedural knowledge, which is knowledge that guides action and decision-making but typically lies outside the scope of consciousness (see also Reber, 1993, pp. 14–17). Unconscious perception tends to lead to automatic and uncontrolled reactions; conscious perception allows individuals to modify their reactions and respond more flexibly to a situation (Merikle & Daneman, 1996, 1998). According to Reber (1993), "the study of unconscious processes generally and implicit learning specifically should be cast into an evolutionary setting" (p. 79). Allowing evolutionary biology to act as an explanatory vehicle for understanding implicit, unconscious mentation and for differentiating these covert processes from explicit, conscious processes may be provocative to some social scientists. However, there is ample evidence (see Reber, 1993; Squire, 1986) that implicit, nonreflective, procedural, or unconscious functions (e.g., procedural memory) are, in terms of evolution, much older, more robust, and less age-dependent than explicit, reflective, declarative, or conscious functions. Infants are able to learn about their social, cultural, familial, physical, and linguistic environments without support from conscious strategies for acquisition (Reber, 1993, p. 97). Why should theoretical concepts on the relations between environment and action, on orientation in space, on local and regional identities, and on cultural memories not include consideration of the psychological and neurological research on implicit learning, implicit memory, and procedural knowledge? Why should one not ask the question of the extent to which the environment does contribute to the development of knowledge? However, arguments about the importance of implicit and explicit knowledge should avoid the "polarity fallacy" (Reber, 1993, pp. 23, 68). Implicit and explicit knowledge should not be treated as though they were completely separate and independent processes. They should instead be seen as interactive components or "as complementary and cooperative functional systems that act to provide us with information about the world within which we function" (Reber, 1993, p. 24).

Advantages of a Spatial Perspective

According to Massey (1999b), spatiality displays the "contemporaneity of difference" (p. 35). The detection, visualization, and analysis of difference are basic tools of any research:

> Space is the sphere of the possibility of the existence of multiplicity; it is the sphere in which distinct trajectories coexist; it is the sphere of the possibility of the existence of more than one voice. Without space, no multiplicity; without multiplicity, no space ... Multiplicity and space are co-constitutive. (p. 28)
>
> The very possibility of any serious recognition of multiplicity and difference itself depends on a recognition of spatiality. (p. 30)
>
> In order for there to be co-existing, multiple histories, there must be space. (p. 35)

Is the new focus of social sciences on spatiality and spatial patterns a relapse into old-fashioned geodeterminism or spatial science? Not at all. In the 1960s spatial patterns were seen as a factor of explanation, and geographers were searching for spatial laws and expected causalities between spatial patterns and actions. Since the 1980s, spatial patterns in most cases no longer serve as an explanation; space is no longer a cause or determining power of human actions. Instead, spatial patterns are perceived as a primary component or focus of cognitive processes. According to Abel (2004, pp. 303–304; see also his chapter in this volume), both information and knowledge are bounded by signs and interpretations. Contents and forms of knowledge cannot be specified, nor can they exist independent of the forms, practices, and dynamics of their underlying systems of signs and interpretation. Observing, classifying, and interpreting *spatial* patterns of signs, objects, relations, flows, and processes are a key to orientation and problem-solving and a means of heuristic exploration. Many situational analyses and decisions demand an ability to draw conclusions from positioning in space or to reconstruct a picture from a small number of signs, clues, or fragments. The ability to recognize, read, and interpret patterns is highly significant, not only for orientation and survival in an unknown or risky environment but also for daily problem-solving and research in many disciplines. Because "nature does not speak" (Ancori et al., 2000, p. 263), the stimuli and signs of the environment have to be perceived, interpreted, and categorized by the knowledgeable agent. In the course of evolution human beings had to learn how to reduce the complexity of spatially ordered signs by promptly recognizing a picture, pattern, entity, context, or gestalt. It is not the sum of a given space's objects or actors that displays social structures and processes but rather the spatial arrangement of and the relations between these objects and actors.

Information perceived by humans is always fragmentary and ambiguous, so it can be interpreted in different ways. Because the search for information cannot go on indefinitely and because an excess of information could even detract from knowledge, humans are constrained by limitations on time, experience, resources, cognitive abilities, attention, and motivation when making inferences about unknown features of their world (see Gigerenzer, 2001). Simon (1956, pp. 129–130) pointed out that there are two sides to bounded rationality: the "cognitive limitations" and, as the title of the article states, the "structure of environments." Gigerenzer (2001) elaborated this notion and stated elsewhere that humans "do not need to wait until all knowledge is acquired and all truth is known ... Adaptive solutions can be found with little knowledge" (Gigerenzer & Selten, 2001, p. 10) if the solutions have to work only in a specific environment. Humans are not supposed to be able to explain the world but to find ways to attain their goals successfully. Interpreting spatial patterns of a given environment helps one understand or describe a situation and recognize ways to solve a problem. For Gigerenzer (2001), ecological rationality is a basic tool of decision-making:

> The notions of psychological plausibility and ecological rationality suggest two routes to the study of bounded rationality. The quest for psychological plausibility suggests looking into the mind, that is, taking account of what we know about cognition and emotion in order to understand decisions and behavior. Ecological rationality, in contrast, suggests looking outside the mind, at the structure of environments, to understand what is inside the mind. These research strategies are complementary, like digging a tunnel from two sides. (p. 39)

He points out that the "rationality" of domain-specific heuristics does not lie in optimization, omniscience, or consistency. Their success is rather in their "degree of adaptation to the structure of environments both physical and social" (p. 38).

This ecological rationality clearly depends to a large extent on the ability to grasp and interpret patterns and entities. It is well known from research on optical illusions that the brains we humans have supplement incomplete information with the help of earlier experiences, prior knowledge, preconceptions, or expectations of behavior (Merikle & Daneman, 1996, 1998; Perrig et al., 1993; Schwan, 2003). The structures we recognize in such patterns and the conclusions we draw from them depend on our prior knowledge, or *Vorverständnis* (Gadamer, 1987a, b, 1960), which means more than cognitive capabilities. It also includes earlier learning processes and experiences, intuition, situational expectations, and the symbolic significance we assign to positioning, goods, buildings, or spatial configurations. Prior knowledge can be defined as a cognitive structure of relationships between signs, events, actions, experiences, memories, and emotions that is possible to retrieve and superimpose on subsequent activities. Choo (1996) used a similar definition for his term "historical knowledge." This retrieval can happen as an unconscious event (e.g., recognition of a face or building), as routines based on former learning processes (e.g., the riding of a bicycle), or as an intentional, conscious act.

A medical doctor has learned how to interpret the image of X-rays. A geomorphologist has been trained to interpret the sequence, stratification, thickness, and spatial arrangement of different types of sediments and remains of organic material in order to gain an insight into climatic conditions and geomorphic processes that took place ten thousand years ago. An archaeologist's task is to reconstruct social structures, power relations, burial rites, and spatial interaction by interpreting the spatial position of stones, ceramics, bones, and other artifacts. To a human geographer thematic maps can serve as a very powerful means of representing, visualizing, and interpreting social structures and processes. Analyzing and interpreting the spatial variation of social indicators on a thematic map is an important heuristic method and can reveal socioeconomic structures, processes, and factors of influence not to be recognized by applying aspatial approaches. However, persons untrained in the relevant discipline might not recognize any structures at all or might not be able to interpret them. Many aspects of society, culture, and economic activity cannot be perceived, described, and explained adequately if the spatial dimension is ignored. The consequences of disregarding the spatial dimension of social structures, indicators, relations, and processes are as adverse as those of neglecting the time dimension or history of social phenomena. Various lines of argument support this assertion. First, both in traditional and modern societies, authority structures, representations of power, distinctiveness, and differences in rank or status are to a large degree spatially exhibited through ordering, positioning, demarcation, exclusion, and elevation. Canter (1991) explained this phenomenon convincingly by pointing out the need for social rules in all human societies (see also Maran, 2006, p. 12). The significance of spatiality for social hierarchies and social relations is also supported by the fact that social ranking is frequently described with spatial metaphors and terminology such as *center*, *periphery*, *top*, *marginal*, *upper* and *lower* class, *insider* and *outsider*, *segregation*, or *distance*.

Space is a means of intervention that controls, manipulates, or otherwise influences the activities of individuals and social systems (see, for example, Feldman, 1997, p. 944; Foucault, 1972, 1980; Townley, 1993). Categorizing, organizing, and commanding space and controlling the spatial arrangement of persons, objects, resources, and relations are very effective devices for governing social systems and manipulating people. "The capacity to dominate and control people or things comes through the geographic location, built-form, and symbolic meanings of a place" (Gieryn, 2000, p. 475). "Space is both the medium and the message of domination and subordination … It tells you where you are and it puts you there" (Keith & Pile,

1993, p. 37). Architectural space constitutes one of the key elements of the symbolism of power. "Social practice always takes place in an environment mirroring the microcosm of social and cultural norms of a given society at a certain time" (Maran, 2006, p. 12). The architecture of even the earliest temples, palaces, and cities distinguished between inside and outside, between the private and public sectors, between holy districts and profane ones, and between areas for upper classes and those for lower classes.

Choreographing space through recurring rituals and ceremonies and through vertical elevation and horizontal distances serves the visualization and confirmation of status, dignity, and prestige (Hölscher, 2006; Weddigen, 2006). It helps a community to recognize, practice and memorize social structures; to strengthen the awareness of hierarchies and dichotomies between us and them, inside and outside, and good and evil; and to reinforce memories and beliefs. In his *Book of Ceremonies* (written about 1488, published in 1516), Piccolomini (1965) devotes an entire chapter to the complex of admittance to and exclusion from the papal chapel. The place of each member of the Curia assembled on the other side of the marble *cancellata* (the place of the pope) was determined by his duties and privileges, all of which were minutely described (see also Weddigen, 2006, p. 272).

"Hardly any other artifact is as closely linked to the human body as architecture" (Juwig, 2006, p. 207). Places, built environments and other materialized spatial structures enable, guide, and constrain action, they arrange patterns of face-to-face interaction that constitute network formation and collective action. They stabilize social life; give structure to social institutions, durability to social networks, and persistence to behavioral patterns (Gieryn, 2002c, p. 35); and facilitate sociality, which may provide the serendipity for new knowledge encounters (Amin & Cohendet, 2004, p. 67). Built environments embody and secure otherwise intangible cultural norms, identities, memories, and values. Built places give material form to the ineffable or invisible, providing a durable legible architectural aide-mémoire (Gieryn, 2000, pp. 473, 481). According to Rapoport (1982), Hölscher (2006), and Maran (2006, p. 12), the built environment can be looked upon as a teaching medium. "Once learned, [the built environment] becomes a mnemonic device reminding us of appropriate behavior" (Rapoport, 1982, p. 67). In the fields of social geography and human ecology, however, controversy abounds regarding the ways in which sociomaterial things can act on humans. One should always bear in mind that the significance of the built environment and architecture "reveals itself only in combination with people and their agency" (Maran, 2006, p. 13) and, as I would like to add, in combination with their prior knowledge, experience, motives, and expectations. Experts in geography, archaeology, and other comparable disciplines are in the position of a detective. In most cases they cannot observe agency; they derive their limited information about social interaction and social structures from surviving clues and objects whose spatial pattern and former meanings they are specially trained to decode and reconstruct.

A second line of reasoning that buttresses the assertion about the negative consequences of disregarding the spatial dimension is that places, monuments, architecture, and built environments are associated with events, histories, biographical experiences, and practices. On the basis of their symbolic meaning, performative spaces create identities, loyalties, and social connectivity; build memories; evoke emotions; and influence feelings. Connerton (1989), Wright (2006), and others argue that the process of remembering in a social sense requires the bodily practice of commemoration in the form of ritual performances. Buildings, courts, mortuary facilities, and streets "facilitate commemorative performance by reproducing and producing social relations" (Wright, 2006, p. 50). Place attachment results from interactive and culturally shared processes of endowing buildings, neighborhoods, or cities with an emotional and symbolic meaning or moral judgment. In cognitive processes, places can function as mnemonic aids, as triggers for emotions and memories, as "spatial anchors for historical traditions" (Foote et al., 2000, p. 305), or as "contextual memory" (Chun & Jiang, 2003). Like icons of power, mnemonic places (Zerubavel, 1997) are specifically designed and constructed to evoke memories, embody histories, and focus the attention of the public on certain objects and interpretations. The more unintentional or unconscious these learning processes are, the more efficient the manipulation of knowledge is.

A third group of arguments proposes that different spatial contexts, environments, and infrastructures offer dissimilar challenges and incentives for learning, research, and problem-solving. "Knowledge cannot be regarded independently from the process through which it is obtained" (Ancori et al., 2000, p. 281). "Intellectual production is always materialized through human bodies, and non-human objects ... Scientists are not faceless organs of scientific rationality, but real people with particular kinds of bodies, histories, skills and interests that make a difference to the kind of knowledge produced" (Barnes, 2004, p. 570). New ideas emerge from social practice, and practice is always undertaken in particular places (Shapin, 1998). "Intellectual inquiry is not the view from nowhere, but the view from somewhere" (Barnes, 2004, p. 568; see also Shapin, 1998). Different places present distinct opportunities of learning and pressures of adapting. They set off different cognitive processes and motivations, induce different discourses, questions, and answers and foster different experiments, practices, and engagements.

Places of discovery can have an impact on scientific results. In various disciplines the process of discovery is not based on formal logic alone; it does not require specific logical methods. Instead, it may involve historical, psychological, and sociological reasoning and research (Hoyningen-Huene, 1987, p. 505). A number of disciplines sample their data in the field, in archives, or in museums. The processes through which they attain their knowledge are highly place dependent (see Wenger, 1998). Different scientific institutions, laboratories, museums, or other places offer different opportunities of learning. They are confronted by different degrees of competition and critique and provide access to different scientific instruments, infrastructure, and resources essential for research. Different departments are integrated into distinct international networks, alliances, and loyalties. They recruit their research staff and visiting scholars from different cultural areas and scientific backgrounds (Jöns, 2003, 2007; Meusburger, 1990; Weick, 1995).

They offer different prospects and risks, and their scholars differ in their scientific biographies, experience, and reputation. Places vary with regard to social control, limitation of research (e.g., stem cells), and the significance of political correctness. The reputation of a research institution may crumble when the alliances and networks associated with a certain theoretical approach falter (Barnes, 2004, p. 588). Different research institutions have different "styles of scientific reasoning" (Hacking, 1985, 2002). For Hacking a style of reasoning connotes both the historio-cultural nature of intellectual projects and their particular nature based upon specialized vocabularies, logics, practices, and forms of explanation. The Japanese notion of *ba* (field) also belongs to this concept. Contrary to kinship, *ba* is a shared space of relationships and mutual commitments built at the place of work. *Ba* is a place in which knowledge is shared, created, and utilized. It is a shared context in cognition and action (Nonaka, 1994; Nonaka & Takeuchi, 1995; Nonaka et al., 2000, p. 8).

The acceptance and reputation of scientific results depend, to a large degree, on where they were generated and verified (Knorr-Cetina, 1992, 1999; Livingstone, 2003; Shapin, 2001; Withers, 2002, 2004). The platform on which scientific results are first presented is often of more importance for their fast spatial diffusion than is the quality or originality of the findings. According to Noteboom (2000) all forms of thought develop out of active interaction with the physical and social environment. In this context scientific practice can be regarded as a process of building networks between actors, resources, things, objects, infrastructures, and social interests (Jöns, 2006, p. 563). If it is accepted that people know and understand through the practice of acting and that acting is always context dependent (Amin & Cohendet, 2004, p. 64), then all forms of learning can be seen as contextual. In geographies of knowledge and in evolutionary economics and organization theory, the external environment of the social system is seen as the driving force that shapes the core competencies, learning processes, and architecture of social systems (see Amin & Cohendet, 2004, p. 57; Geser, 1983; Meusburger, 1998; Mintzberg, 1979). Reviewing the relevant literature in economics of knowledge, Amin and Cohendet (2004) draw the conclusion "that the powers of context—spatial and temporal—should be placed at the center of any theorization of knowledge formation" (p. 86).

Another advantage of using the spatial dimension for perceiving and displaying social phenomena, structures, and processes lies in the fact that space can be represented and visualized on various scales. Each spatial scale (i.e., level of aggregation and generalization) exposes different structures and patterns not visible or not clearly perceivable on other scales. An overload of information on the microscale may blur patterns that are quite clear at the meso- or macroscale where information is reduced to the most important elements. Each scale enables distinctive insights, heuristic assumptions, and interpretations hardly possible on another scale. Different scales put forward different research questions and may call for different theoretical approaches. Maps of various scales may function as "knowledge mediaries" or "active knowledge actants" (Amin & Cohendet, 2004, p. 71).

How Is It Possible that Sociomaterial Things Positioned in Space Act upon Humans?

Material environments which provide cultural meaning (e.g., in the form of action settings) can order social relationships and the course of activities. The symbolic meaning of a place or action setting determines what is regarded as appropriate behavior (see Weichhart, 2003). Conduct tolerated backstage or in private may not be appropriate or permissible in public. It depends on the categorization and demarcation of places and areas whether individual action (e.g., spraying graffiti or parking a car) or collective action (e.g., a demonstration) is illegal and how the police respond to it. It goes without saying that action settings (such as a mosque, synagogue, a chemist's laboratory, or a cinema) do not *determine* the behavior of people. Rather, they prompt people to act in a particular way that is appropriate for the cultural significance of the place. People knowing which behavior is appropriate, permitted, tolerated, desired, or disapproved of in a particular action setting behave in a certain way. Individuals who are not aware of the symbolic or cultural significance of an action setting do not behave in accordance with the expectations.

Do places or built environments have an impact on action per se, apart from their symbolic meaning and apart from powerful people or organizations occupying them? Or is it only their symbolic meanings that influence human action? How do sociomaterial things positioned in space act upon humans? These questions are part of an intense debate in a number of disciplines. Diverse answers are offered by actor-network theory (Jöns, 2003, 2006, p. 563; Latour, 1987), human ecology (Weichhart, 2003), symbolic action theory (Boesch, 1991), science studies (Gans, 2002; Gieryn, 2000, 2001, 2002a, c; Goss, 1988; Livingstone, 2003; Withers, 2002, 2004), and subject-centered action theory (Werlen, 1993, 1995, 1997b). Authors following the traditional path of sociology do not acknowledge the agentic capacity of material realities. Durkheim's classical notion that the social cannot be explained by the material and that "the truths of science are independent of any local context" (Durkheim, 1899/1972; Gieryn, 2002c, p. 45) is still widely accepted in sociology. However, Durkheim himself used spatial disparities as an analytical tool in his book on suicide (Durkheim, 1897/1997). Advocates of actor-network theory and modern science studies have no difficulties with regarding built environments as constitutive—along with governance structures, legal processes, and workplace organization (Gieryn, 2002a, p. 343).

As Gieryn (2002c, pp. 37–38) has elaborated in detail, Giddens (1984a, b) is disinclined to ascribe autonomous agency to built environments and instead makes them a function of interpretations and uses by knowledgeable humans. He is reluctant to allow that buildings or spatial structures may preempt or preclude the agent's conscious apprehension, interpretation, or mobilization and that they can structure practices without necessarily requiring actors' cognizant involvement. Giddens supports the idea that "location is only socially relevant—and this is crucial—when filtered through frames of reference that orient individuals' conduct" (see preface in Werlen, 1993, p. xv). Werlen is equally unwilling to attribute agency to material objects. He accepts the constraining character of material artifacts, but maintains that such objects are "always and only constituted and reconstituted through subjective agency" (Werlen, 1993, p. 199). Bourdieu (1989) does not share Giddens's reluctance. For him, buildings become "objectified history: systems of classification, hierarchies, and oppositions inscribed in the durability of wood, mud, and brick" (Gieryn, 2002c, p. 39; see also Bourdieu, 1981, pp. 305–306). Gieryn (2002a) summarized the debate in sociology as follows:

> Once upon a time, sociologists thought that the effects of "the social" (political or economic interests, power, face-sheet attributes, discursive forms, etc.) on scientists' legitimate beliefs about the natural world were limited to the institutional contexts for problem choice, data collection, experimentation, publication, funding, or peer review. The content of what would become scientific truth was determined by the given reality of the natural world; social factors just introduced error or governed the pace at which nature revealed its secrets. Then came revolution number one: scientific truth became a social construction, and the race was on how to show how the content of scientific claims was substantially (completely?) affected by power, interests, discourse.... The "natural world" itself dissolved into so many representations or accounts, and reality became the upshot of persuasion and negotiation (losing its force as a cause of belief). Then came revolution number two, inspired by the slow realization that it didn't make sense to leave reality out of truth making. But "nature" was brought back in not as antipode to "the social" (as it was at the beginning) but as part of it. Nowadays, ... social things and natural things have autonomous force in shaping scientists' beliefs and practices. "Given reality" has an effect on the content of claims and theories, but only as that stuff is suspended in vast networks of circulation, along with people, meanings, political interests, economic power, and too many other things to list. Neither nonhuman physical reality nor human social reality can be privileged as an explanation or cause of what scientists believe or write. (p. 341)

For sociologists of science, the era of human or social omnipotence is over. Posthumanist sociology (Latour, 2001; Knorr-Cetina, 2001; Pickering, 1995) redistributes agency among diverse causal powers—human, material, social, ideational (Gieryn, 2002a, p. 342). Recently, sociology has become interested in the "significance of material culture in social life" (Gieryn, 2000, p. 465). Social processes (difference, power, inequality, collective action) happen *through* the material forms that humans design, build, use, and protest (Gieryn, 2000; Habraken, 1998). The culturally reproduced images of places are arbitrary in their social construction but real in their consequences—for what people do consciously or routinely (Gieryn, 2000, p. 473). As with any generalization, there are always some exceptions to the main trend. Werlen drew my attention to Linde (1972), who recognized the relevance of "real things" for sociology long before it became fashionable.

Allen (1977), Galison (1997, 1999), Gieryn (2002c), Knorr-Cetina (1992), Livingstone (2003), and others have tried to answer the question of whether and why architectural layouts of offices and laboratories do have effects on the generation of scientific results and the performance of scientists. Empirical evidence from many studies suggests that the architecture of buildings and the floor plan of laboratories have effects on patterns of social interaction among scientists, on casual face-to-face contact and chance encounters among those scientists working on different projects or in different teams (Gieryn, 2002c, pp. 46–47). "Arrangements of space inside research laboratories reproduce the divisions of labor and even status hierarchies among a discipline's practitioners" (p. 47). But as physical environments can express social meaning by acting as a system of signs, they matter for science in a semiotic sense as well (Hillier & Hanson, 1984, p. 8). When new scientific fields emerge, the architecture of laboratories has to be changed. "Campus buildings originally designed to house biology here, chemistry there, and physics down the street now become impediments to biotechnical research that demands practitioners, skills, and equipment from all three disciplines" (Gieryn, 2002c, p. 50).

When Did Scientific Interest in the Spatiality of Science, Knowledge, and Education Evolve?

Scientific and political interest in spatial disparities of knowledge (literacy, research, educational attainment, educational infrastructure, and professional skills) harks back to the first decades of the 19th century. It was the time when social reformers in France and the United Kingdom believed that poverty, crime, and alcoholism were caused by ignorance and a lack of moral education and when relations between knowledge and economic performance were discovered. In the 19th century, scholars in the social survey movement studied social and spatial disparities of illiteracy, the availability and quality of schools, the skills and salaries of teachers, the availability of books in households, and the educational attainment of children (see Furet & Ozouf, 1977; Heffernan, 1988, 1989; Meusburger, 1998, pp. 191–198). In 1826, C. Dupin gave a lecture about the interrelation between the population's educational achievement and economic well-being. In 1827, he published the *Carte figurative de l'instruction populaire de la France*, a map that depicted large regional disparities in educational attainment between northern and southern France. The tables that were added to that document compared the educational attainment, the number of patents for inventions, and the membership in the *Académie Française*, with various economic indicators suggesting a correlation between educational achievement and economic performance. To my knowledge, the first map of spatial disparities of education on a global scale was published by Alexander von Humboldt on the topic of *geistige Bildung* (intellectual and spiritual culture) (Berghaus, 1838–1848/2004, p. 143). Fletcher (1849) published a map on "Ignorance in England and Wales" (reprinted in Hoyler, 1996, p. 188).

In the decades thereafter, academics in the social sciences and the humanities were occasionally interested in the relations between knowledge, space, and place. Since the 1960s, however, research on spatial disparities of knowledge, science, technology, and education has increased remarkably in a number of disciplines. The geography of knowledge and education emerged in German speaking-countries in the early 1960s (Geipel, 1965, 1968, 1969, 1971). Some of the main research issues of geography of education between 1965 and 2007 were spatial disparities of educational achievement (Geipel, 1971; Meusburger, 1980), location criteria and catchment areas of educational institutions (Kramer, 1993), the spatial distribution of jobs for the highly and marginally educated work forces, the relation between the hierarchy of a national urban system and the educational achievement of the workforce (Meusburger, 1978, 1980, 1996b, 2000, 2001b), the relation between spatial mobility and educational achievement (Meusburger, 1980), ethnicity and educational achievement (Frantz, 1994; Freytag, 2003; Gamerith, 2006; Meusburger, 1996a), spatial disparities in the feminization of the teaching profession (Schmude, 1988), provenance, and the careers and mobility of scientists (Beaverstock, 1996; Jöns, 2003; Meusburger, 1990; Weick, 1995). Research reports about the geography of education have been presented by Meusburger (1976, 1980, 1998, 2001a), and Butler & Hamnett (2007).

In the 1960s and 1970s studies on the diffusion of information, the role of face-to-face contact, and the location of offices and headquarters contributed substantially to knowledge about why jobs of highly skilled decision-makers and experts tend toward spatial concentration and clustering (Goddard, 1971; Goddard & Morris, 1976; Goddard & Pye, 1977; Hägerstrand, 1966; Hägerstrand & Kuklinski, 1971; Meusburger, 1978, 1980; Pred, 1973; Thorngren, 1970; Törnqvist, 1968, 1970; Westaway, 1974). Seminal influence on economic geography came from the theory of human capital (Schultz, 1960, 1963) and from research on innovations and innovative firms (Feldman, 1994; Feldman & Florida, 1994; Kline & Rosenberg, 1986; Lundvall, 1988; Sternberg, 2007) and inventions (Nelson, 1959a, b, 1962), which have been seen as the most important sources of competitive advantage and as the driving force of economic development since Schumpeter (1912). They were followed by studies on the role of institutions in regional development (Camagni, 1991; Maskell & Malmberg, 1999; Storper, 1995; Storper & Venables, 2004), the relations between technology and economic development (Malecki, 1980, 1997, 2000), learning economies, regions and cities (Gertler, 2003; Lundvall, 1997; Maskell & Malmberg, 1999; Matthiesen, 2004; Morgan, 1997), collective learning (Capello, 1999; Keeble & Wilkinson, 1999; Lawson & Lorenz, 1999; Stam & Wever, 2003), learning organizations (Maskell & Malmberg, 1999), knowledge creation (Ibert, 2007), knowledge-creating companies (Nonaka, 1994; Nonaka & Takeuchi, 1995), and industrial clusters (Malmberg & Maskell, 2002; for an overview see Bathelt et al., 2004).

Another line of research on knowledge and power was the role of travel accounts and geographical imaginations in the production of imperial knowledge (Gregory, 1994, 1998, 2000; Pratt, 1992) and the relationship between power and knowledge in the conduct of former and present colonialism (Gregory, 2004).

The geography of science, which has developed since the early 1980s mainly in the United Kingdom and the United States (Livingstone, 1987, 1995, 2000, 2002, 2003; Naylor, 2002, 2005a, b; Ophir & Shapin, 1991; Powell, 2007; Shapin, 1988, 1991; Shapin & Schaffer, 1985; Withers, 2002, 2004), had epistemic roots other than the geography of knowledge and education. The notion that place matters in the production of scientific knowledge began to take shape in the 1930s, when Fleck (1935/1980) pointed out that the question of what is regarded as "scientific fact" depends on *Denkstilen* (styles of thinking) and *Denkkollektiven* (collectives of thinkers) or *Denkgemeinschaften* (communities of thinkers). "Even the simplest act of observation is conditioned by thinking style and is, hence, tied to a community of thinkers" (Fleck, 1935/1980, p. 129).[2] Hayek (1937, 1945) distinguished between context-specific knowledge, which he called knowledge of the particular circumstances of time and place, and knowledge of general rules, which he called scientific knowledge. Kuhn (1962) elaborated similar ideas. In the 1970s a number of historians and sociologists of science questioned whether there was an inherent universality of scientific content. They argued that knowledge reflects various social interests of those who propose it (Bloor, 1976), that science is a particular kind of social practice, that scientific results are socially constructed (Latour, 1987; Latour & Wolgar, 1979), and that they reflect unequal relations of power and uneven distribution of resources (Barnes, 1998, p. 205; Jöns, 2006, p. 562). In this debate Latour (1987) reminded the scientific community that "the proof race [of the sciences] is so expensive that only a few people, nations, institutions, or professions are able to sustain it, this means that the production of facts and artifacts will not occur everywhere and for free, but will occur only at restricted places at particular times" (p. 179). His concept of cycles of accumulation in scientific centers of calculation describes the way in which certain places can become centers that dominate the periphery: "At every run of this accumulation cycle, more elements are gathered in the center ... at every run the asymmetry between the foreigners and the natives grows" (p. 179). "Systematic knowledge is never free of context and prescriptive assumptions. Hence, each group will make knowledge claims according to its interests and strategic goals. Integration of knowledge is based on rhetoric, persuasion skills, and power rather than established rules of discovering the truth" (Renn, 2001, p. 13651). As soon as it was accepted by most social scientists that the generation of scientific knowledge is situated in time and space and that truth about natural reality is influenced by the social environment (Haraway, 1988; Knorr-Cetina, 1992, 1999; Kuhn, 1962; Latour, 1987; Schaffer, 1991; Shapin, 1998; Shapin & Schaffer, 1985), new research questions about the meaning of space within the process of knowledge production arose and paved the way for a geography of science (Jöns, 2006, p. 561; Livingstone, 2003).

Major stimuli for a spatial turn in science studies originated partly with those historians and sociologists of science who shifted their research focus from problems of truth and validity to issues surrounding the credibility of and trust in scientific experiments and the circulation of scientific results (Ophir & Shapin, 1991; Schaffer, 1991; Shapin, 2001). The spatial turn was also facilitated by researchers who switched from producing laboratory ethnographies that focused on the local aspects of science practice to viewing the laboratory as cultural space (Naylor, 2002, 2005a, b; Schaffer, 1998; Shapin & Schaffer, 1985). According to Powell (2007), "due to a concern for the credibility of truth-claims

[2] Auch das einfachste Beobachten [ist] denkstilbedingt, also an eine Denkgemeinschaft gebunden.

and truth-claimants, science studies *necessarily* had to confront questions of spatiality" (p. 310).

Naylor (2005b, p. 3) distinguishes between three geographies of science. The first one is the microgeography of science focusing on the spaces (e.g., laboratories) in which scientists have done their work. The second one is a consideration of science and its contexts, including the city, the region, and the nation. The third geography is focused on a more general and abstract concept of the relation between science and the public sphere. National censuses, national academies of science, ordnance surveys, and other enterprises have been used to construct national identity and unity (Naylor, 2005b, p. 8).

Shapin (1988, p. 373) showed how, in the 17th century, the siting of knowledge-making practices contributed to the credibility of experiments. Truth-claims of scientific experiments needed spaces such as laboratories where witnessing was to occur and could be guaranteed by a community of respected scholars. Other sites of knowledge generation and legitimation were museums, archives, lecture halls, botany gardens, and selected field sites. Such sites acted as "truth spots" (Gieryn, 2002b) facilitating experiments and practices, bringing certain actors together and excluding others, and legitimating results (for detailed discussions see Gieryn, 2002a, b, c; Knorr-Cetina, 1992; Livingstone, 2003; Naylor, 2005a, b; Ophir & Shapin, 1991; Powell, 2007; Schaffer, 1998; Shapin, 1988, 1991; Shapin & Schaffer, 1985).

Geographers of science are interested in all steps of the generation, dissemination, and application of knowledge. They study the settings in which scientific experiments and studies were carried out and the places where scientific knowledge was generated, displayed, and legitimated. According to Livingstone (2003), science "is a human enterprise situated in time and space, ... scientific knowledge bears the imprints of its location" (p. 13). He has pointed out that "space matters in the conduct of scientific inquiry" (Livingstone, 2002, p. 8) and that "in different spaces different kinds of science are practiced" (Livingstone, 2003, p. 15). He has described distinctive geographies of writing and reception (p. 29), showing that the generation of knowledge requires a spatial context other than the showing of experiments and that the legitimation of scientific results, in turn, calls for other locations:

> A gulf thus opens up between what was called the "trying" of an experiment and the "showing" of an experiment ... The shift from "trying" to "showing," from delving to demonstrating ... is a spatial manifestation of the move from the context of scientific discovery to the context of justification. (p. 24)

The distinction between the context of discovery and the context of justification reaches back to the mid-19th century (for details see Hoyningen-Huene, 1987, pp. 502–503) and was already a central theme of the *Wiener Kreis* (Carnap et al., 1929), of Popper (1934), and other authors. Hoyningen-Huene (1987, p. 508) suggests a differentiation that is at least threefold. In the first phase a theoretical idea, a hypothesis, or a theory sketch is "generated." This process may be initiated by a challenge, a problem to be solved, a discourse, or the crossing of disciplinary borders. In the second phase the plausibility of the idea is assessed. Finally, the elaborated idea may be subjected to critical testing and, if it is successful, it may be "accepted." The criteria or communal cognitive values involved in this testing vary both in the spatial and the temporal dimensions.

Other important stimuli came from psychology in the 1980s, when learning and creativity were no longer regarded as mere cognitive processes of individuals but as something influenced by interaction with social and cultural contexts and artifacts, especially by participation in cultural activities. As soon as psychologists saw the learning of individuals in relation to social systems, contexts, networks, interactions, and social practices, as soon as it was accepted that action settings, situations, or a system's environment can influence the creation, diffusion, and application of new knowledge, social and environmental psychologists had built a bridge to human geography.

In economics the boom in publications on the role of knowledge in economic performance, on learning organizations, on the formation and distribution of knowledge in firms, on knowledge formation in clusters, on innovative milieus and other issues started mainly in the 1990s (Aydalot, 1986; Aydalot & Keeble, 1991; Camagni, 1991, 1995; Christensen & Drejer, 2005; Lam, 2000; Lorenzen & Maskell, 2004; Maillat et al., 1993), although many classics (e.g., A. Smith, S. Mill, L. Stein, L. Walras, A. Marshall, J. Schumpeter, S. Kuznets, and F. Hayek) had pointed out that knowledge and innovations are the key driving force of economic development (for details see Meusburger, 1998, pp. 81–96; Nelson, 1959a, b, 1962; Nelson & Winter, 1977). According to Amin and Cohendet (2004, p. 17) traditional economists had to overcome at least four theoretical obstacles before knowledge could become central in economic theory. They had to (a) abandon the "vision of knowledge as a simple stock resulting from the accumulation of information in a linear process," (b) shed "the hypothesis that any form of knowledge can be made codifiable," (c) give up "the vision that knowledge is limited to individuals", and (d) "the idea that knowledge is limited to something that people 'possess'." I add, that they had also to accept that place and spatiality matter.

Reading the literature on networks and clusters, one gets the impression that many authors take it for granted that networks and clusters contribute almost automatically to the generation of knowledge (see Bathelt & Glückler, 2000; Bathelt et al., 2004; Lo, 2003; Schamp & Lo, 2003). My view is that networks and clusters per se have no positive effects on the generation of knowledge, they can even detract from the generation and transfer of important knowledge. Whether networks generate new knowledge depends on who belongs to the network, how much expertise the network comprises, which interests the members of the network pursue, and how links are added and removed. A proper understanding of most networks requires that analysts characterize the assembly process that generated them, that they increase their knowledge about the structure of collaboration and about the ways in which people form alliances (Barabási, 2005, p. 640).

Forces and Processes Generating and Reproducing Spatial Disparities of Knowledge

Power, Knowledge, and the Organization of Space

Among the primary causes of spatial disparities of knowledge, the most prominent are the division of labor, the growth of complex social systems, the emergence of hierarchies, and the asymmetry of power relations in social systems. The vertical division of labor implies that a profession or activity (e.g., the production of a shoe or machine) formerly performed by a single person is broken down into various activities carried out by many individuals with different levels of skills and decision-making authority. Some lines of routine work become deskilled and need less training. Other activities (e.g., research, design, and marketing) require high-level, time-consuming training and call for specialized expertise and skills. The bifurcation of skills means that jobs of highly skilled professionals and high-ranking decision-makers shift to the top levels of an organization's hierarchy, whereas low-skill routine activities in production and administration are predominantly located at the lower levels of the hierarchy. In the spatial dimension this process leads to the emergence of centers and peripheries of different ranks. Positions of power and authority and highly skilled experts show a strong tendency toward spatial concentration in a few centers, whereas low-skill routine activities coordinated and controlled by external decision-makers show a trend toward dispersion and decentralization (Meusburger, 1996b, 1998, 2000, 2001b).

Any invention or new technique that facilitates indirect communication over large distances also enlarges the potential for a spatial division of labor, improves the opportunities of governing and coordinating large organizations in space, intensifies the coalition between knowledge and power, encourages the growth of cities, and reinforces the disparities of knowledge between the center and the periphery. Since the close coalition between knowledge and power and their dialectical relationship may be regarded as the main reason for the long persistence and continuous reproduction of spatial disparities of socioeconomic structures, the questions arise as to why knowledge and power depend on each other, why they mutually transform each other (Brown, 1993, p. 154), and why their top ranks tend toward spatial proximity.

The importance of power to the production and dissemination of certain types of knowledge can hardly be overestimated. Since early history, it has been in the interest of those in power to control or influence institutions of knowledge production. Also "in modern societies the ability to facilitate or suppress knowledge is in large part what makes one party more powerful than another" (Flyvbjerg, 1998, p. 36). Political and cultural elites fake documents, invent "facts" (e.g., the existence of weapons of mass destruction in Iraq) and construct historical memories that legitimate their actions and provide national or regional identities. The ways we know history are determined more by contemporary concerns of those in power than by history itself (Williams, 1973, p. 9).

In order to obtain power and preserve it for notable periods of time in an uncertain, risky, and dynamic social environment, a social system has to be successful in achieving (and redefining) its goals and has to retain its ability to learn and adapt to a dynamic environment. In a dynamic and competitive society, the acquisition of knowledge and skills is a process that never reaches completion. The skills and knowledge needed for the key functions of a social system striving for success (i.e., survival) will always be scarce and expensive. The larger the uncertainties, the greater the social demand to *anticipate* prospective events and future developments or to reveal a hidden truth. This pressure leads to emergence of oracles, dream readers, priests, advisors, experts, intellectuals, and think tanks, which derive their privileges and status from their claim to know better or earlier than the majority of people or to represent a link to the mysterious and unrevealed.

The relation between knowledge and power has been discussed intensely by a large number of philosophers and social scientists (Foucault, 1972, 1980; Konrád & Szelényi, 1978; Mann, 1986; Meusburger, 1998; Nietzsche, 1888; Stehr, 1994a, b; Weber, 1978). If rulers of empires or high-ranking decision-makers of large social systems want to maintain their power, survive competition, preserve their legitimacy, and impose their view of the world, they need the support of two types of experts. First, they depend on the analytical skills and professional competence of experts of analytical knowledge. Second, they need the support or assent of the representatives of orientation knowledge, which was called *Heilswissen* by Scheler (1926), to legitimate their power. With regard to the single actor, it is clear that both categories are strongly interrelated and influence each other. However, on the level of organizations, a clear functional differentiation and specialization can be observed. Experts of analytical knowledge have other tasks, need different training and skills, and use other methods than experts of orientation knowledge do (Meusburger, 2005).

Analytical knowledge, scientific knowledge, competence, and proficiency are needed in order to analyze a situation as precisely as possible and to offer solutions to problems that have to be solved. Experts are persons who, by objective standards and over time, consistently show superior and outstanding performance in typical activities of a particular domain (Gruber, 2001). The gaining of expertise is usually characterized cognitively "as a process of enhancing one's competence in a target domain by accumulating experience of problem solving, understanding, and task performance in that domain" (Hatano & Oura, 2001, pp. 3173–3174). Experts are needed and paid to predict the likely consequences of actions, to anticipate potential opportunities and risks and to give advice on how to cope with uncertainties. They are supposed to reduce complexity and offer more certainty in a risky environment than a layperson is able to do. They are required to anticipate, perceive, and understand new developments and offer solutions to new-found problems and challenges. They are expected to interpret signs and patterns of change that are not understood by most people. As expertise is action-orientated advice, it should be free of errors. The role of an expert involves that person's trustworthiness, accountability, and credibility. "Trusting becomes the crucial strategy to deal with an uncertain, unpredictable, and uncontrollable future" (Sztompka, 2001, p. 15913; see also Sztompka, 1999).

Because "bodies of expert knowledge ... are widely taken as the touchstone of truth in our culture" (Shapin, 2001, p. 15926), credibility is the most important asset of an expert. The relation between the expert and the layperson but also that between experts of various domains can be described as "epistemic dependency" (Jones, 2001, p. 15917). Incompetence, ignorance, and lack of experience are important factors leading to the collapse of social systems or to the decline of centers. Therefore, those in power depend on the analytical capabilities and competence of experts.

However, it is not sufficient just to acquire power; power has also to be legitimated. Rulers achieve the legitimization of their power mainly from representatives of orientation knowledge. In earlier times they were prophets or oracles; later they were priests, intellectuals, editors, propaganda departments, novelists, and artists. Orientation knowledge provides a point of reference, declares what is good or evil, bestows identity and forms the glue that keeps a social system together. Representatives of orientation knowledge are trained and experienced in the art of influencing, convincing, and manipulating people. Their task is not to analyze a "real situation" or to search for truth or "objective facts" but rather to sustain the internal cohesion and motivation of their social system, to create beliefs and collective memories, to mobilize loyalty, to justify actions, and to make moral judgments.

Through the mechanism of moral exclusion, the dichotomy between "good" and "evil" is equated with "us" and "them." The specialists of orientation knowledge are responsible for depicting their "own side" as representative of moral values, justice, peace, and human rights, as acting upon God's wishes or being "God's own country" (Weinberg, 1935). The opposing side is demonized as an aggressor, a barbaric enemy, a danger to peace, a power of darkness, an axis of evil, or a war criminal (Jewett & Lawrence, 2003; Lawrence & Jewett, 2002; Wunder, 2006). The mechanism of moral exclusion is not a modern invention; it has been used since ancient times (Assmann, 2000; Meusburger, 2005; Wunder, 2005). In most cases, moral exclusion was combined with spatial exclusion. The enemy or barbarian was outside, or had to be excluded from the community.

In a conflict, the representatives of orientation knowledge define whether a person is a terrorist or a freedom fighter, a hero or a war criminal. Their tasks might include supporting the propaganda of their government or party, glorifying or demonizing historic events, manipulating or censoring media, falsifying documents, or constructing new "collective memories." The party who succeeds in imposing their definitions, interpretations, and memories is already well on the way to winning the conflict. Therefore, opponents do everything they can to achieve hegemony in the interpretation of texts, the definition and explanation of historical facts, the construction of narratives, and the use of images and symbols.

In periods of conflict, however, it is not easy to keep a balance between the two categories of knowledge. Orientation knowledge can cloud perception, prevent a realistic assessment of situations, foster prejudice and chauvinism, and lead to decisions that trigger damaging consequences for the stability of the social system. More than a few governments, political parties, and organizations have failed to reach their goals because they took their own propaganda and myths as reality and were no longer able to evaluate a situation and foresee the consequences of their actions.

The Architecture of Social Systems and the Location of Knowledge

The survival chances, competitiveness, or success of large and complex organizations depend to a large extent on the questions of how competence, expertise, and high-level decision-making authority are allocated within the social system, how formal hierarchies and communication structures are ordered, and how spatially allotted specialized knowledge is coordinated. In this context, the term *hierarchy* is not defined as a top-to-bottom chain of command in which all levels differ from each other in their degrees of authority and privileges. Instead, hierarchy is defined as a functional differentiation of a complex system. Once an organization attains a certain size and complexity, it cannot exist without adopting hierarchic structures of communication, information-processing, and decision-making. According to Simon (1962) and Reber (1993, 75) "evolutionary useful" systems are virtually always hierarchical. The main purpose of a hierarchy is to reduce complexity and uncertainty, to increase the number of information channels to the environment, and to improve the organization's ability to acquire, transfer, and exploit knowledge effectively.

Ultimately, an organization can compensate for only a certain amount of incompetence, so it acts in its own interest when it fills the key positions of information-processing, decision-making, planning, coordination, and control with knowledgeable and skilled persons. In particular, those positions and subsystems that are constantly confronted with high degrees of uncertainty and whose decisions have enduring consequences for the entire system require highly skilled and experienced decision-makers. In social systems knowledge, skills, and experience have the same function as redundancy in technical systems. They reduce uncertainty and enhance survival chances in a dynamic and risky environment.

Because important or valuable knowledge is always scarce, the first crucial question is where to locate scarce knowledge, important skills, and high levels of decision-making within the architecture of an organization. The architecture of a social system (its structure of information-processing, formal communication, and decision-making), is not a matter of deliberate choice. The optimal architecture of an organization depends on the goals of the organization; the degree of uncertainty confronting it; the constancy or instability of its environment; the system's autonomy, size, and complexity, and the available instruments and channels of information-processing (Geser, 1983; Meusburger, 1980, 1998; Mintzberg, 1979). In systems with stable goals and low degrees of uncertainty, decision-making, problem-solving, research, development, and planning shift to the upper levels of the system's hierarchy, with the lower levels predominantly retaining routine activities and jobs for marginally skilled workers. This arrangement is typical of a bureaucratic organization. In systems dealing with a dynamic and complex environment and with constantly changing, unpredictable, one-time transactions, decentralization of competence and authority within the organization is more effective than such centralization (Mintzberg, 1979).

The second question is where to locate scarce knowledge, important skills, and high-ranking decision-making in the spatial dimension. Most large and complex social systems are not autonomous and free in their choice of where to locate their highest levels of authority, decision-making, and knowledge production. From the viewpoint of organization theory, it is again primarily the degree of uncertainty with which a decision-maker must cope that decides the optimal location of a position. The fiercer the competition and the greater the uncertainty about the consequences of far-reaching decisions, about future developments, and about the correctness of methods and objectives, the more necessary it is to have frequent, spontaneous, face-to-face contact with knowledgeable, well-informed, high-ranking decision-makers and highly skilled specialists of *other* organizations and other domains.

Uncertainty can be temporarily reduced through constant and prompt acquisition of specialized knowledge of important innovations, future technical and economic developments, and probable societal changes. Continuous acquisition of new knowledge and early access to crucial information make it possible to adapt quickly to new situations and to cope with new challenges. Early information and new knowledge about important developments are no guarantee for successful actions; indeed, they tend to provoke new questions and new uncertainties. But a continuous search for information and knowledge increases a social system's transparency, predictability, efficiency, and competitiveness, at least for a while.

This kind of crucial information is not presented in the Internet, business reports, press conferences, public data bases, or scientific journals. It is first revealed by rumors, through nonverbal communication in informal meetings, and in small fragments that have to be pieced together like a puzzle by the attentive observer. Few centers or nodes of network-building offer potential for high-ranking, spontaneous, face-to-face contact with top decision-makers of other institutions. Gaining access to informal interest groups, prestigious clubs, and powerful networks offering this kind of early, exclusive, and valuable knowledge is a matter of mutual trust. If trust is not founded on kinship, it has to be earned and maintained by frequent face-to-face contact, conditioning of moods and sentiments through rituals, affinity of interests, empathy, and a record of mutually useful performance (Brown, 1993; Glückler & Armbrüster, 2003). Trustworthiness can also be achieved by membership in prestigious institutions, by living or working at the "right" address or by belonging to networks of high reputation. Trust in the reliability of partners and in the superior knowledge of experts is an indispensable prerequisite for coping with an uncertain, unpredictable, and uncontrollable future (Sztompka, 2001, p. 15913). Mutual trust cannot be established by telecommunication. The generation of trust is tied to places. It develops by common practice, symbolic acts, ceremonials, and rituals that require copresence in certain secluded or distinguished places.

It is not only the functional role of face-to-face contact but also the symbolic meaning and reputation of a place that attract high-level decision-makers, intellectuals, and other successful knowledge producers. Authenticity, credibility, accountability, and trustworthiness are in many cases associated with the symbolic meaning of certain places or territories. HipHop musicians (Mager, 2007) are not the only people who derive their authenticity, credibility, and reputation partly from the places they are associated with or belong to; so do bankers, lawyers, scientists, actors, and members of other professions. Places are a kind of acronym of the complexity of a social system, historical event, or economic structure. Acronyms help individuals cope with the overload of information they are exposed to. Since the information-processing capabilities of an individual are limited and because that person cannot check and process all the detailed information needed for successful action, people constantly work with simplifications, generalizations, and cognitive reductions. Symbols or names of places stand for complex institutions, situations, and actions. Harvard stands for a prestigious university with thousands of prominent scholars and students. New York's 47th street is a symbol for expertise and reputation in the trade of diamonds. Zürs and Davos stand for expensive jet-set skiing; Hollywood, for media power. Other places may be associated with war crimes, torture, or danger.

Each large and complex organization displays its asymmetric power relationships, its functional division of labor, and its structures of decision-making and coordination in a spatial hierarchy of places. The center or core of a social system, economic sector, or scientific discipline is defined as the place where the highest degree of authority is located (Gottmann, 1980; Meusburger, 1980, 2000, 2001b). Centrality is the spatial manifestation of power, authority, and prestige. In early civilizations, the center was a sacred place where the connection with superhuman beings was initiated. Sages and priests were assembled at the center of power or presented themselves as the center of the social system. By virtue of their connection with their gods, forebears, or other superhuman beings, they claimed preeminence with regard to authority, knowledge, and competence and represented divine and ancestral will in everyday life. Similar ritualistic constructions of centers exist in modern societies as well.

A center is the nodal point of interaction and communication from where the elements of a social or spatial system are governed, coordinated, and controlled (Strassoldo, 1980). A center is a point of reference and orientation. It collects and distributes resources and sets the rules, norms, and standards for the members of the system. A center legitimates knowledge. It offers more diverse and wide-ranging knowledge sources, early access to crucial information, and a higher potential for high-level face-to-face contact than less important places. Centers derive some of their attractiveness through their national and global connectivity with other centers of knowledge and power. Through the business connections of big corporations and institutions, centers are able to absorb vastly diverse kinds of knowledge from elsewhere and profit from a wide range of information channels. The concentration of expertise and high-level decision-making, the high degree of connectivity and the consistent generation of new knowledge imbue centers with a special "buzz" (Storper & Venables, 2004) or atmosphere (Böhme, 1998). In economic geography, the term *buzz* initially referred to "the information and communication ecology created by face-to-face contacts, copresence and colocation of people and firms within the same industry and place or region" (Bathelt et al., 2004, p. 38). Contrary to most industrial clusters, the information and communication ecology of high-ranking urban centers is characterized by a large diversity of industries, institutions, cultures, and knowledge bases.

The term *center* or *core* also has a psychological meaning. It is associated with social attributes such as power, authority, dominance, prestige, access to resources, attractiveness, and influence. Most experts, scientists, and intellectuals are fascinated by domain-specific authority or centrality, want to concern themselves with the essentials of existence, and strive for influence and recognition. They are convinced that they have something important to convey to humanity, that their capabilities are needed by society, and that they can offer solutions to important problems. Being associated with a high-ranking center endows experts with prestige and influence. Proximity to power increases their chances of influencing important decision-makers. Someone at the periphery is seen as an "outsider"; he or she has less influence, fewer resources, and less prestige. That person may also be marginalized. Centers act like magnets for highly skilled professionals, experts, scientists, artists, and intellectuals striving for prestige, influence, or success. Because centers and peripheries are socially constructed and because space is a product of relations and interactions and is always "in a process of becoming … never finished … never closed" (Massey, 1999b, p. 28), the rank, significance, and locations of centers and peripheries change over time. In some cases (nomadic tribe, army in war) the location of the center moves constantly.

The recent discussion of face-to-face contact has four weaknesses. First, many authors (e.g., Amin & Cohendet, 2004; Bathelt et al., 2004; Storper & Venables, 2004) do not distinguish between orientation contacts, planning contacts, and routine contacts as was suggested earlier by Goddard (1971), Goddard and Morris (1976), Goddard and Pye (1977), Hohenstein (1971), and Meusburger (1980). From the theoretical point of view, it is not advantageous for the face-to-face contacts of sales girls or clerks to be lumped together in the same category as those of top managers or scientists. Face-to-face contacts of orientation need other locations of learning and have other spatial interactions than face-to-face contacts of planning or routine work. Routine face-to-face contact can be more easily replaced by letters or electronic communication than is the case with face-to-face contact of orientation.

Second, the need for interagent face-to-face contact and the relevance of proximity undergo a kind of life cycle during the relationship of the people involved. In the first phase, when interactions have to be established and the degree of uncertainty is high, face-to-face contact may be extremely important. When the agents come to trust each other, much face-to-face contact can be replaced by electronic communication, and proximity loses importance.

This lack of distinction between types of contact, levels of management and expertise, and degrees of uncertainty has led to the third weakness, an overemphasis of proximity and clusters in what is know as "new regionalism" (see also Kröcher, 2007, pp. 57–61). There is no general "proximity imperative" (Lagendijk, 2001, p. 146) in human geography. The questions are: proximity to whom, to which purpose, and for which reasons? Need for proximity to reduce transport costs (e.g., within production and supply chains of industrial clusters) should be distinguished from need for proximity to learn from and imitate successful agents and competitors (so as to reduce uncertainty). It should also be distinguished from a need for proximity to benefit for symbolic reasons (e.g., to gain reputation and trust by belonging to a center of authority). If an organization is highly autonomous (e.g., as a global market leader) and enjoys a stable environment with little or no competition (e.g., public administration), or if it can enhance its reputation in ways other than identification with centers of domain-specific authority, then proximity to other institutions is almost irrelevant.

The fourth problematic trend in the research on clusters and industrial districts is the overemphasis on homogeneous business cultures and on in-group relations between persons, companies, and institutions already known to each other (either as a supplier or a competitor) and in more or less regular mutual communication. According to Porter (2001) "clusters are geographic concentrations of interconnected companies, specialized suppliers, and service providers, firms in related industries; and associated institutions … in particular fields that compete but also cooperate" (p. 144). Creativity hardly develops in homogeneous business cultures. It emerges by drawing analogies from completely different domains that previously had nothing to do with one another. Creativity is very often based on transgressing boundaries. Combinatorial creativity requires a rich store of knowledge and the ability to form links between many different types of knowledge.

Cultural Hegemony, Cultural Areas, and Clashes of Orientation Knowledge

Epistemic hegemony is a means of domination and a capacity to control and manipulate people (Brown, 1993, pp. 154, 164). The filtering of information and the manipulation of attention are effective tools for exercising power. Long-term hegemonic filtering or manipulation of information clearly creates areas where certain topics or contents of knowledge prevail while others are suppressed or criticized. One can easily define areas of political prejudice, barefaced lying, cultural and historical ignorance, bigotry and racism, and flourishing conspiracy theories. Cultural hegemony is an attempt to determine which religions, ideologies, values, traditions, collective memories, narratives, and interpretations of historical events should be accepted or tolerated in its area of influence and which should be rejected. In extreme cases power rests on the principle of *cuius regio eius religio* ("whose the region, his the religion"), the proviso by which the religion of the sovereign is automatically that of all the subjects as well. Political, economic, and cultural elites produce public sentiments and stereotypes with the help of media, educational institutions, and other channels of communication and invent traditions and historical memories that legitimate their actions and support national or regional identities. Governments try hard to preserve the image of the good country. "Behavior inconsistent with the defensive, clean, law-abiding, faithful, and humble stance demanded by the stereotype must be denied or hidden" (Jewett & Lawrence, 2003, p. 231). Conflicts are often portrayed as a dichotomy between good versus evil, a fight between right versus wrong, human dignity and freedom versus tyranny and oppression (for details see Jewett & Lawrence, 2003; Lawrence & Jewett, 2002; Meusburger, 2007). In some cases elites go so far as to maintain that their nation has a manifest destiny (Weinberg, 1935), that it is God's own country, or that it has God's chosen people. If "the enemy is demonic and the saints are perfectly pure, no matter what they may do in battle" (Jewett & Lawrence, 2003, p. 222), any aggression and torture seems to be justified.

The relations between culture and behavior as well as between knowledge and action are ambiguous and heavily disputed. However, most authors would agree that culture shapes aspirations, stereotypes, understanding, ways of learning, frames of interpretation, and collective memories. Epistemological cultural relativism goes so far as to claim that culture determines what we humans know and how we know it. According to Herskovits (1948), reality is perceived through the spectacles of culture. He asserts that all human experience of the physical world as well as of society is culturally mediated. In his theory, all perceptions, evaluations, and judgments are a function of the cultural system to which one belongs (Harouel, 2001, p. 3181).

If knowledge can be understood as adequately justified true interpretation based on and determined by a system of signs and interpretation (Abel, 1999, pp. 304–310; see also his chapter in this volume), then it can easily be manipulated if those in power are successful in changing the system of signs and interpretation. In disputes the distinction between opinion, belief, and knowledge becomes blurred and irrelevant. Believing is as effective a disposition or capacity to act as knowing is. Subjectively binding beliefs suffice for action; a person who believes in something is prepared to accept the consequences for his or her actions (for more details see Abel, 2004, pp. 161–169).

Media, schools, museums, and other cultural institutions are deployed by power elites to generate, disseminate, and support a particular set of beliefs and orientation knowledge and to transmit their culture and collective memories from one generation to the next. Cultural institutions are supposed to enforce collective beliefs (memories that support the ideology and goals of the dominant political elite) and to ignore or suppress other narratives. Striking examples are national centennial celebrations of revolutions, civil wars, and, in immigrant nations such as the United States or Australia, glorious formative moments that often ignore the history of natives and various immigrant groups (see Spillman, 1998).

Because cultural knowledge is created through practices, interaction, and social control at particular places, schools are not only a place of instruction or formal education. They are also a site and context where social relations evolve and where identities, self-awareness, goals, beliefs, attitudes, cultural preferences, discourses, stereotypes, and social inequalities are produced or reproduced and where parents, teachers, and other role models interact. In multiethnic states, schools have been considered the main instrument in educating "backward" or "uncivilized" ethnic minorities. If the state authority or a dominant political party controls the contents of textbooks or the recruitment of teachers, it can direct students' attention to certain issues, divert it from others, eliminate a large number of possible interpretations, and destroy ethnic self-confidence of minorities. However, this attempt is resisted by people belonging to other cultures and subcultures with contradicting memories and interpretations of the world. In order to secure the survival of their culture, ethnic or religious minorities strive to organize learning opportunities for their young (Hatano & Takahashi, 2001, p. 3041). This response is one of the reasons why the public school system in multiethnic states has often been a focal point of power struggles, an arena where cultural conflicts and clashes of knowledge are the most intense (Frantz, 1994; Freytag, 2003; Gamerith, 2006; Meusburger, 1996a, 2003; Tomiak, 1991; Trueba, 1989). In modern society hegemony is not necessarily expressed by suppression or censorship but a shift of public attention to certain issues.

Most clashes of knowledge have a spatial dimension, at least for a certain span of time. Cultural space is defined as an area or set of places in which certain kinds of orientation knowledge are considered true or correct by the power elites or the majority of the resident population and where the collective orientation knowledge is bolstered and legitimated by traditions, practices, and cultural artifacts. The extent, visibility, degree of homogeneity, and consistency of cultural areas vary over time. They partly depend on the ability of elites to control collective memories, organize consent and support among followers, construct and interpret "realities," influence collective knowledge and actions, mobilize solidarity and a sense of belonging, and mark places or territories with their cultural artifacts (e.g., signs, flags, monuments, street names, and styles of architecture). The purpose of such activity is to guide the collective knowledge and memory of the respective population in a certain direction and to erase other events from the memory of future generations. Throughout history, changes of ruling dynasties or political systems have coincided with iconoclashes (Foote et al., 2000; Hoyler & Jöns, 2005; King, 1997; Latour, 2002).

Apart from science studies and geography of religion, very few human geographers have discussed the spatial dimension and spatiality of orientation knowledge or ideology. One of them is Sahr (2006), who identified in Brazil a hierarchical space of traditional Roman Catholicism, a communitarian and syncretic space of rural ideologies, an individualistic approach of Protestantism, a rhizomatic space of Afro-Brazilian religions, and a fluid space of Amerindian religions, all of them partly counteracting, through social actions, the imposed modernist development ideology of the nation-state.

When dealing with cultural space, one must avoid the "territorial trap" (Agnew, 1999). The territorial trap is entered into when it is assumed that all actors within a culturally defined area behave in a similar way or follow the same norms. It would be wrong to assume that culturally defined space, for example, is devoid of opposition, divergent behavior, conflicts, social differentiation, or social change. It is important to emphasize that the concept of culture does not imply homogeneity in cultural consciousness or practice within an ethnic group, cultural category, or area. Each ethnic group and cultural area has its internal differentiation and conflicts, its elites and subcultures. Culture is not a stable system of signs and interpretations but rather a process and place in constant motion, where meaning and situated identities connected to ethnicity, language, or religion are continuously created and performed (Bellwood, 2001; Wunder, 2005). Members of ethnic or cultural groups continuously borrow and adopt new cultural forms and alter their identities through contact-induced learning. Being rooted in a culture does not mean immunity against new ideas, norms, or practices. Instead, it suggests that agency and intentionality are bounded by a certain tradition of meanings and values that differ from that of other cultures. The assertion that hegemonic manipulation of information creates cultural areas never means that the whole population is thinking or acting in a certain way. However, mapping and interpreting the *spatial* distribution of ideas (e.g., Darwinism, enlightenment, creationism), performances (e.g., Mozart's itineraries and the career paths and mobility of scientists), and artifacts (e.g., the distribution of baroque churches in Europe) or analyzing *spatial* disparities in the predominance of narratives, norms, opinions, or public discourses as represented in media or opinion polls can be an important heuristic device in the research processes of the humanities and social sciences. How many people orient their actions to these narratives and norms is another question altogether. Culturally defined spaces or spatial arrangements of cultural artifacts are not something fixed or self-contained. They are constantly changing, negotiated, and contested.

Spatial Diffusion and Mobility of Knowledge: An Attempt to Construct a More Realistic Communication Model

The most efficient way to transfer rare or specialized knowledge from place A to place B is through the migration of those people who dispose of that knowledge. However, they will only be as successful in place B as in place A if they find comparable conditions in B. All other attempts at knowledge transfer, such as the sending of texts, construction plans, instruments, and machines are no guarantee that the knowledge is fully understood or accepted by the receivers. Striking evidence of this notion was witnessed at the end of World War II, when the American, Russian, and British Forces were eager to obtain the most advanced technological and scientific information Germany had to offer. The U.S. Commerce Department's Office of Technical Services addressed the industries of the United States with the following words: "[Y]our government is offering you a chance to share in the war's reparations—reparations in the form of technological information— ... in all fields of industry and research [including] testing methods, chemical research, new products, new materials, production methods, and plant development" (as cited in

Gimbel, 1990, p. 57). According to Gimbel (1990), 4,994 Allied investigators in Germany microfilmed millions of pages of patent applications, construction plans, and research results. However, as the Office of Technical Services had to admit in December 1947, "it has been to our experience that the worthwhile developments cannot be exploited successfully or without considerable expense unless the German technicians familiar with all the details of such developments are brought to this country" (as cited in Gimbel, 1990, p. 57). Finally, 765 leading German scientists and engineers were brought to the United States in Operation Paperclip, and 350 rocket technicians in Operation Overcast. Other scientists were taken to the Soviet Union.

The process of knowledge transfer from person X (producer of knowledge or sender of information) in place A to person Y (recipient or potential user of information) in place B is a very challenging and complex research issue. The speed at which new knowledge diffuses through a spatial system depends on many factors. Just a few of them are the type of knowledge, its usefulness to power, its relevance for economic competition, the institution within which the new knowledge is produced, the competence of the producer in articulating or codifying his knowledge, the interest of the producer (inventor) in sharing his or her knowledge, the prior knowledge necessary to understand the substance of new information, the availability of technology necessary for the production and application of knowledge, and the inclination to accept and use the knowledge.

Following the example of Arrow (1962), Machlup (1962), Nelson (1959a, b), and others, an entire generation of economists treated scientific and technological knowledge as information (for more detailed discussion see Ancori et al., 2000, p. 256; Cowan et al., 2000, p. 221). Some social scientists and philosophers (e.g., Spinner, 1994) did so as well. Most economists recognize the existence of tacit knowledge but restrict their analysis to codified knowledge, which, in their opinion, can be reduced to information that is easily transferable to other decision agents. Ancori et al. (2000) explained why the codification of knowledge is a major concern of economists and why they find it difficult to give up their claim that there is almost no difference between codified knowledge and information. To be treated as an economic good with discernible and measurable characteristics, knowledge must be put into a form that can be exchanged, and that form is information. This view has been challenged not only by sociologists of science (Callon et al., 1999; Collins, 1983, 1985; Stehr & Meja, 2005), geography (Livingstone, 1995, 2005; Meusburger, 1998), and philosophy (Abel, 2004) but recently also by economists (Amin & Cohendet, 2004; Ancori et al., 2000; Cohendet & Meyer-Krahmer, 2001; Dosi & Marengo, 1994; Pavitt, 1998).

The diffusion of knowledge cannot be reduced to the mere transmission of information. Unlike information, which is very mobile and can spread all over the world in seconds, knowledge is rooted in persons, institutions, routines, and regional cultures. From the viewpoint of the producer of new knowledge or *sender* of a message, the boundary between information and knowledge might become blurred. In regard to the recipient of a message, the difference between knowledge and information becomes quite distinct. As soon as spatial dissemination of knowledge becomes an issue, a distinction between knowledge and information and between different types of knowledge becomes indispensable.

However, it is not sufficient to distinguish between tacit and codified knowledge (Polanyi, 1958) or between declarative and procedural knowledge (Anderson, 1983). The terms *tacit* and *codified* knowledge can be accepted as the opposite ends of a continuum, but these categories are fluid. It is not possible to draw a generally valid line between tacit and codified knowledge. What is tacit knowledge for one person or at one point in time can be perfectly explicit for other actors or at some other time. Knowledge may remain tacit just because the emitter and receiver have no knowledge about how to exchange knowledge (Ancori et al., 2000, pp. 273–274; Baumard, 1999; Collins, 2001). Some authors view codified and tacit knowledge as essentially complementary because all forms of codified knowledge require tacit knowledge to be useful (Ancori et al., 2000, p. 257). Cowan et al. (2000, p. 213) criticized "that the terminology and meaning of 'tacitness' in the economics literature [have] drifted far from its original epistemological and psychological moorings [and have] become unproductively amorphous." Some authors confuse tacit knowledge with nonverbal knowledge. According to Abel "tacit knowledge means those aspects of knowing that are implicit in situations of perceiving, speaking, thinking and acting, but are not made explicit, are not disclosed at [the] surface" (Abel, 2004, p. 322; see also his chapter in this book). Tacit knowledge must be distinguished from nonverbal knowledge (e.g., the competence at playing the violin) that cannot be articulated by using linguistic expressions. Although the concept of tacit knowledge is widely discussed (see Ancori et al., 2000; Baumard, 1999; Collins, 2001; Cowan et al., 2000; Gertler, 2003; Lam, 2000; Polanyi, 1967, 1985; Reber, 1993), most publications do not distinguish between implicit and nonverbal knowledge but rather treat them synonymously.

In mainstream economics, too much emphasis has been put on the producer and codifier of knowledge. It is important to keep in mind that successful codification does not imply automatically that the codified knowledge will be widely disseminated. From the viewpoint of geography, increased emphasis should be put on the recipients of information and on the factors that influence the communication process between the producers of knowledge (senders of information) and the receivers of information. The quality and accuracy of codifying knowledge is only one side of the coin. The other side is that of the cognitive abilities, orientation knowledge, interests, motivation, attention, emotions, and prejudices of the recipients of information. The producers and transmitters of knowledge have limited influence on the extent to which their knowledge is accepted or interpreted elsewhere. A certain type or content of knowledge may be perfectly codified in equations, published in international journals, and well understood by 50 to 100 theoretical physicists, but the rest of the world population may just not have acquired the prior knowledge necessary to read and understand the mathematical equations and apply them to its benefit. Therefore, I question the assumption shared by Fujita et al. (1999), Maskell and Malmberg (1999), and many others that the more codified the knowledge involved, the more mobile it is and that knowledge, once codified, is almost instantly available to all firms at zero cost regardless of their location.

In order to better understand the complexity of the communication process between person X in place A and person Y in place B, I propose a communication model pertaining to only a small selection of processes intervening between the producer of knowledge (the sender of information) and the receiver of information (Fig. 2.1). Depending on the type of knowledge and the topic under investigation, this model could and should be greatly elaborated and amended by a number of further issues, such as the questions of how knowledge is legitimated, how individual knowledge becomes collective knowledge, how knowledge is transformed into routines and organizational structures, and how the communication process is influenced by an organization's size and hierarchy.

The communication process displayed in Fig. 2.1 consists of nine stages: (a) the willingness of person X to share his or her knowledge with others, (b) the ability of person X to verbalize and codify that knowledge, (c) the degree of attention, reputation, and visibility of the platform where the information is presented, (d) the code in which a message is written, (e) the communication channel used for transmission, (f) the chances of a recipient to receive the information, (g) the ability of the receiver to read the used code, (h) the prior knowledge of the receiver to understand the information and integrate it into his or her knowledge base, and (i) the willingness

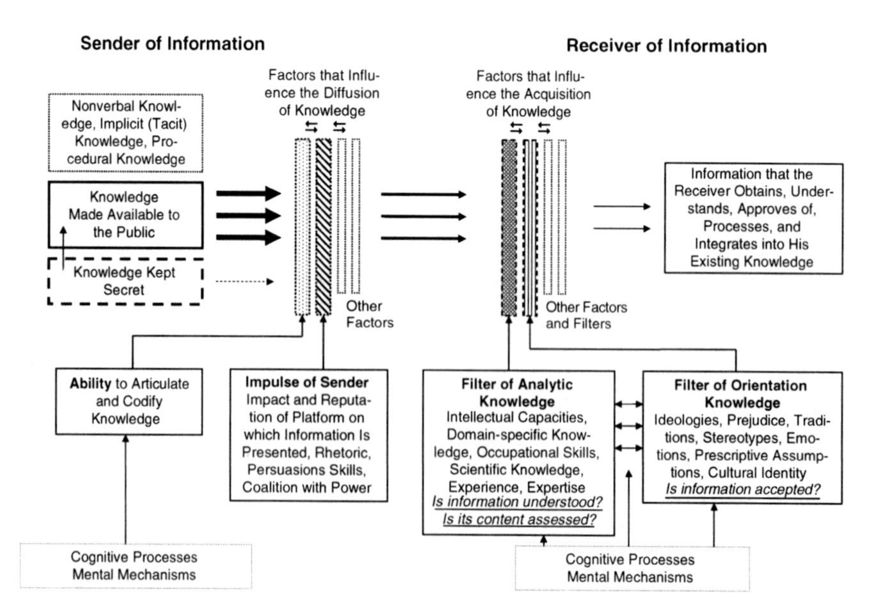

Fig. 2.1 "Factors influencing the transfer of knowledge between persons at different places" (P. Meusburger. With permission)

of the receiver to accept the new information. Each stage of the communication process has a high degree of actor-, community- and place-dependent contingency and acts like a filter, letting some information pass and withholding or transforming other information. At each step and place of the communication process, there can be misunderstandings, distortions, misrepresentations, and loss of information that may result in further spatial disparities of knowledge.

Any visualized model runs the risk of being misunderstood as a description of static relations and mechanistic interactions. In reality, these processes and steps of communication are not arranged sequentially as depicted in Fig. 2.1. They must be conceived of as interactive learning loops that incorporate agents, individual and collective capabilities, work practices, spatial and organizational contexts, resources, and strategic visions. The terms *prior knowledge*, *analytical knowledge*, and *orientation knowledge* are not understood as a static knowledge base but rather as a knowledge base subject to a continuous process of change. Prior knowledge is not something people possess; it is something they constantly develop.

As two filters, the receiver's analytical knowledge and orientation knowledge should not be viewed as separate and unconnected. They are related to each other and influence each other in many ways. Orientation knowledge may motivate a person or social system to acquire new scientific knowledge, but it can also distort perception, weaken analytical judgment, and prevent the scientific investigation of topics—with possibly unpleasant results. The acquisition of new analytical or scientific knowledge may contribute to the revision of prejudice, stereotypes, and ideologies.

The first step in a communication process concerns the question of whether a producer of new knowledge is *willing* to share his or her knowledge. Knowledge that improves chances and competitiveness, promises high profits, or constitutes the role of an expert is in many cases kept secret as long as possible. In many situations it may be an advantage to leave competitors or opponents uncertain about one's goals and actions. A new bargain is normally made public only after it has been signed. Some scientific results may be shared only after they have been patented. The act of keeping knowledge secret, or restricting access to it, has a long tradition. Many religions have holy knowledge that priests or shamans pass on only to chosen successors or have temple precincts and sanctums that only priests are allowed to enter. Worldwide, billions of dollars are spent to prevent industrial or military espionage.

The second question affecting the communication is whether a producer of knowledge is *able* to codify his or her knowledge to express it in language, signs, and gestures or to transform it into physical objects (e.g., scientific instruments). Each person knows more than he or she is able to articulate to someone else. The producer of knowledge has to transform ideas and matter into language or signs "in order to generate comprehensible and well-communicable scientific claims about much more complex phenomena" (Jöns, 2006, p. 570). During each transformation from matter to sign, there is not only a loss of multiplicity, particularity, locality, and materiality but also a gain of standardization, compatibility, relative universality, and immateriality (see Jöns, 2006, p. 571; Latour, 1999, pp. 70–71).

Different producers of knowledge are proficient in different codes. Some of the codes are understood by a large number of people; others, by only a few. A manuscript published in Estonian has far fewer potential readers than a publication in English. However, the message in Estonian may be much more important or deserve a wider distribution than that in English.

The third factor that can enhance or confine dispersion of knowledge concerns the platforms where new knowledge is presented. Experts, scientists, professionals, and artists require a platform of attention that puts them in the spotlight and guarantees their presence in the relevant media. Different platforms send impulses of varying strength, have dissimilar reputation, visibility, and audibility and achieve unequal attention. Because human memory and information-processing capacities are limited, attention is selective and limited (Franck, 1998). The selectivity in perception determines what is learned and kept in memory and what is excluded. Judgment of significance is neither impartial nor spatially invariant; it is an instrument for exercising power. Considering today's flood of information, the contents of a message or its usefulness for society are often less important for its wide diffusion than the platform on which it is presented. The locality where new knowledge is proclaimed determines to a large extent the relevance, visibility, and credibility of the knowledge claims and the attention of the media. Channels of transmission (e.g., books, journals, radio, TV, Internet, and congresses) differ in their reach, credibility, and effectiveness.

On the side of the receiver, incoming information has to pass at least two filters before it is processed. In this context, the term *filter* is a metaphor for various cognitive processes and factors that influence the selectivity of perception, the evaluation and interpretation of incoming information, and the conversion of knowledge into practice. The fact that somebody has access to a piece of information does not mean that he or she is interested in it; understands its meaning; reflects upon it; recognizes its far-reaching implications; can associatively link the piece of information with his or her existing structure of knowledge; or accepts the information as relevant, valid, or credible. The perception, interpretation, evaluation, and acceptance of information requires more or less extensive or specialized prior knowledge, which cannot be transferred easily from one person to the next.

The first filter consists of domain-specific knowledge and expertise, the familiarity with codes (foreign languages, mathematical equations), and various cognitive abilities, such as the skills of analyzing problems or evaluating situations. This filter decides whether the recipient is able to find, read, and understand the message; evaluate the importance of the information correctly; integrate it into his or her knowledge base; and transform the knowledge into action. Prior knowledge is also indispensable when it comes to coping with the increasing overload of information. The learning processes necessary for acquiring this type of prior knowledge may require notable amounts of time and money. Publications of molecular biology or high-frequency physics are available worldwide after they have been published. However, persons who have not completed years of study in the subject have little or no use for the available information. Some types of scientific knowledge cannot be simply transferred from A to B; they must be replicated in B with expensive experiments in sophisticated laboratories (see also Callon et al., 1999; Collins, 1983, 1985).

The second filter on the side of the recipient of information falls into the category of orientation knowledge. It may consist of religious and ideological convictions; a set of dispositions, prejudices, and stereotypes inculcated in childhood; emotions; national myths; political legends; loyalty to a community; cultural traditions; and so on. This filter determines whether a new piece of information is compatible with the recipient's values and identity. Orientation knowledge decides whether new information is emotionally accepted or rejected. Information may be rejected because it questions the recipient's own cultural identity, integrity, or convictions or because it shatters collective memories, historical myths, or the reputation of the institution a person belongs to.

Both filters on the receiver's side are embedded in contexts and influenced by social processes, the selectivity of communication, interpersonal interaction, social control, circular mobility, value systems, and many other factors. The effects of these filters and others are the most important reason why the dissemination of certain categories and contents of knowledge are limited to certain places and areas and to cultural, religious, and political contexts. The effects also explain why those categories and contents of knowledge circulate only within and between particular areas with similar preconditions and bypass others. The spatial distribution and spatial mobility of those people who can read the relevant codes (e.g., a foreign language or a mathematical equation), who have the prior knowledge to understand the codified message, and who have access to the communication channels and resources needed to apply the codified knowledge deserve much more scientific interest (Jöns, 2007). One can extend the model by including additional factors of influence and filters; by describing institutional, cultural, and political contexts in which the individuals process information; by distinguishing between a language form, a picture form, and an action form of knowledge, as Abel suggests in this book; and by focusing more on the signointerpretational practices. In Abel's view "contents of knowledge and forms of knowledge *cannot* exist independent of the forms, practices, and dynamics of the underlying representational, interpretational, and sign system" (p. 15 in this volume). These signointerpretational practices greatly vary in the spatial dimension and could also be integrated into the model.

With regard to the outcome of the communication process between person X in place A and person Y in place B, as described in this model, knowledge can be differentiated into at least five categories, distinguishable by the speed and places of their diffusion:

1. Knowledge that is kept secret as long as possible or necessary in order to gain a competitive advantage.
2. Knowledge that is widely disseminated in the interest of its producer, though a number of barriers may impede its diffusion (e.g., a sender's difficulty expressing his or her knowledge in language, signs, gestures, or performance, or insufficient attention attracted by the platform on which the knowledge is presented).

3. Knowledge that is successfully codified and publicly available but understood, processed, and applied only by a relatively small epistemic community with the prior knowledge necessary to read the code (e.g., foreign language or mathematical equation) in order to comprehend the message or replicate the experiment.
4. Knowledge that is successfully codified, well documented, open to the public, and well understood by the addressees but not accepted or adopted by a distinct group of recipients for emotional or ideological reasons.
5. "Common knowledge" that is easily articulated and disseminated, easily acquirable, promptly understood, and relatively conflict free, making it the only one of these five categories of knowledge that is as mobile in space and as ubiquitously distributed as hypothesized in traditional economics.

It goes without saying that combinations of these five types also exist.

Conclusion—The Knowledge-Transfer Paradox

The neoclassic contention that codified knowledge is highly mobile may now be refuted in social and economic sciences, but it is still en vogue in research policy and regional policy. Even prestigious scientists, such as R. N. Zare, former chairman of the U.S. National Science Board and currently professor at Stanford University, hold the view that knowledge is ubiquitously available. "This is an age of 'knowledge and distributed intelligence,' in which knowledge is available to anyone, located anywhere, at any time" (Zare, 1997, p. 1047). However, a closer look at those disciplines dealing with knowledge proves the opposite. The issues of sending, receiving, and processing information and of generating and transferring knowledge are studied by anthropologists, archaeologists, brain researchers, computer scientists, economists, geographers, historians, linguists, neuroscientists, philosophers, psychologists, sociologists, and scholars in other disciplines as well. If codified knowledge were really as mobile as some observers maintain, and if knowledge really did diffuse through barter exchange among pairs of agents in communication networks as some economists still assume (e.g., Cowan & Jonard, 2004), why does it take 10 to 20 years or even longer for important scientific results and theoretical concepts to move from one discipline to the next even when they are located in the same university town?

An answer might lie in what I propose to call "the knowledge-transfer paradox." It refers to the fact that some of the scholars who act on the assumption that codified knowledge is very mobile in space and accessible to anyone have not the faintest idea of what other disciplines, epistemic communities, or languages contribute to their own research topic. Even some of the most reputed scientific journals accept manuscripts whose authors were unaware that work on their topic, idea, or concept had been published 10, 20, or 30 years earlier by other scholars in another language or another discipline. In this chapter, however, I have outlined some of the ways in which easy access to information neither guarantees the acceptance and application of available knowledge nor eradicates spatial disparities of knowledge.

Clearly, the study of knowledge production and knowledge transfer as a social construction and a context-dependent practice remains to become "one of the most vibrant and exciting areas of research in the social sciences and humanities" (Thrift et al., 1995, p. 1).

References

Abel, G. (1999). *Sprache, Zeichen, Interpretation* [Language, signs, interpretation]. Frankfurt am Main, Germany: Suhrkamp.
Abel, G. (2004). *Zeichen der Wirklichkeit* [Signs of reality]. Frankfurt am Main, Germany: Suhrkamp.
Agnew, J. (1999). Mapping political power beyond state boundaries: Territory, identity, and movement in world politics. *Millenium, 28*, 499–521.
Allen, T. J. (1977). *Managing the flow of technology*. Cambridge, MA: MIT.
Altvater, E., & Mahnkopf, B. (1996). *Grenzen der Globalisierung. Ökonomie, Ökologie und Politik in der Weltgesellschaft* [Limits of globalization: Economic realities, ecology, and politics in world society]. Münster, Germany: Westfälisches Dampfboot.
Amin, A., & Cohendet, P. (2004). *Architectures of knowledge: Firms, capabilities, and communities*. Oxford, England: Oxford University Press.
Ancori, B., Bureth, A., & Cohendet, P. (2000). The economics of knowledge: The debate about codified and tacit knowledge. *Industrial and Corporate Change, 9*, 255–287.
Anderson, R. (1983). *The architecture of cognition*. Cambridge, MA: Harvard University Press.
Arrow, K. J. (1962). Economic welfare and the allocation of resources for invention. In National Bureau of Economic Research (Ed.), *The rate and direction of inventive activity: Economic and social factors* (pp. 609–625). Princeton, NJ: Princeton University Press.
Assmann, J. (2000). *Herrschaft und Heil. Politische Theologie in Altägypten, Israel und Europa* [Rule and salvation: Political theology in ancient Egypt, Israel, and Europe]. Vienna, Austria: Hanser.
Aydalot, P. (Ed.). (1986). *Milieux innovateurs en Europe* [Innovative milieus in Europe]. Paris: GREMI.
Aydalot, P., & Keeble, D. (Eds.). (1991). *High technology industry and innovative environments in Europe*. London: Routledge.
Barabási, A. L. (2005). Network theory—the emergence of the creative enterprise. *Science, 308*, 639–641.
Barnes, T. J. (1998). A history of regression: Actors, networks, machines, and numbers. *Environment and Planning A, 30*, 203–224.
Barnes, T. J. (2004). Placing ideas: Genius loci, heterotopia and geography's quantitative revolution. *Progress in Human Geography, 28*, 565–595.
Bathelt, H., & Glückler, J. (2000). Netzwerke, Lernen und evolutionäre Regionalentwicklung [Networks, learning, and evolutionary regional development]. *Zeitschrift für Wirtschaftsgeographie, 44*, 167–182.
Bathelt, H., Malmberg, A., & Maskell, P. (2004). Clusters and knowledge: Local buzz, global pipelines and the process of knowledge creation. *Progress in Human Geography, 28*, 31–56.
Baumard, P. (1999). *Tacit knowledge in organizations*. London: Sage.
Beaverstock, J. V. (1996). Migration, knowledge and social interaction: Expatriate labour within investment banks. *Area, 28*, 459–470.
Bellwood, P. (2001). Cultural evolution: Phylogeny versus reticulation. In N. J. Smelser & P. B. Baltes (Eds.), *International encyclopedia of the social & behavioral sciences* (Vol. 5, pp. 3052–3057). Amsterdam, The Netherlands: Elsevier.
Berghaus, H. (2004). *Physikalischer Atlas oder Sammlung von Karten, auf denen die hauptsächlichsten Erscheinungen der anorganischen und organischen Natur nach ihrer geographischen*

Verbreitung und Vertheilung bildlich dargestellt sind. Zu Alexander von Humboldts KOSMOS. Entwurf einer physikalischen Weltbeschreibung [Physical atlas or collection of maps depicting the main manifestations of inorganic and organic nature by their geographic range and distribution]. Frankfurt am Main, Germany: Eichborn. (Original work published 1838–1848).

Blackler, F. (2002). Knowledge, knowledge work, and organizations. In C. W. Choo & N. Bontis (Eds.), *The strategic management of intellectual capital and organizational knowledge* (pp. 47–62). New York, Oxford: Oxford University Press.

Bloor, D. (1976). *Knowledge and social imagery*. London: Routledge & Kegan Paul.

Boesch, E. E. (1991). *Symbolic action theory and cultural psychology*. Berlin: Springer.

Böhme, L. (1998). *Atmosphäre* [Atmosphere]. Frankfurt am Main, Germany: Suhrkamp.

Bourdieu, P. (1981). Men and machines. In K. Knorr-Cetina & A. Cicourel (Eds.), *Advances in social theory and methodology: Toward an integration of micro- and macro-sociologies* (pp. 304–317). London: Routledge & Kegan Paul.

Bourdieu, P. (1989). *Social space and symbolic power. Sociological Theory, 7,* 14–25.

Brown, R. H. (1993) Modern science: Institutionalization of knowledge and rationalization of power. *The Sociological Quarterly, 34*, 153–168.

Brunés, T. (1967). *The secrets of ancient geometry—And its use*. Vol. 1. Copenhagen: Rhodos.

Butler, T., & Hamnett, C. (2007). The geography of education: Introduction. *Urban Studies, 44,* 1161–1174.

Cairncross, F. (1997). *The death of distance: How the communications revolution will change our lives*. Boston, MA: Harvard Business School.

Callon, M., Cohendet, P., Curien N., Dalle, J.M., Eymard-Duvernay, F., & Foray, D. (1999). *Réseau et coordination* [Network and coordination]. Paris: Economica.

Camagni, R. (1991). *Innovation networks: Spatial perspectives*. London: Belhaven.

Camagni, R. (1995). Global network and local milieux: Towards a theory of economic space. In S. Conti, E. Malecki, & P. Oinas (Eds.), *The industrial enterprise and its environment:spatial perspective* (pp. 195–216). Aldershot, England: Avebury.

Canter, D. (1991). Social past and social present: The archaeological dimensions to environmental psychology. In O. Grøn, E. Engelstad, & I. Lindblom (Eds.), *Odense University Studies in History and Social Sciences: Vol. 147. Social space: Human spatial behaviour in dwellings and settlements* (pp. 10–16). Odense, Denmark: Odense University Press.

Capello, R. (1999). Spatial transfer of knowledge in high technology milieux: Learning versus collective learning processes. *Regional Studies, 33*, 353–365.

Carnap, R., Hahn, H., & Neurath O. (1929). *Wissenschaftliche Weltauffassung—Der Wiener Kreis* [Scientific conception of the world—The Viennese Circle]. Vienna: Artur Wolf.

Casey, E. S. (1996). How to get from space to place in a fairly short stretch of time: Phenomenological prolegomena. In S. Feld & K. H. Basso (Eds.), *Senses of place* (pp. 13–52). Santa Fe, NM: School of American Research Press.

Cassirer, E. (1944). *An essay on man*. New Haven, CT: Yale University Press.

Choo, C. W. (1996). The knowing organization: How organizations use information to construct meaning, create knowledge, and make decisions. *International Journal of Information Management, 16*, 329–340.

Christensen, J., & Drejer, I. (2005). The strategic importance of location: Location decisions and the effects of firm location on innovation and knowledge acquisition. *European Planning Studies, 13*, 807–814.

Chun, M. M., & Jiang, Y. (2003). Implicit, long-term spatial contextual memory. *Journal of Experimental Psychology: Learning, Memory, and Cognition, 29*, 224–234.

Cohendet, P., & Meyer-Krahmer, F. (2001). The theoretical and policy implications of knowledge codification. *Research Policy, 30*, 1563–1591.

Collins, H. M. (1983). The sociology of scientific knowledge: Studies of contemporary science. *Annual Review of Sociology, 9*, 265–285.

Collins, H. M. (1985). *Changing order. Replication and induction in scientific practice*. Beverly Hills, CA: Sage.

Collins, H. M. (2001). What is tacit knowledge? In T. Schatzi, K. Knorr Cetina, & E. von Savigny (Eds.), *The practice turn in contemporary theory* (pp. 107–119). London: Routledge.
Connerton, P. (1989). *How societies remember*. Cambridge, England: Cambridge University Press.
Cowan, R., & Jonard, N. (2004). Network structure and the diffusion of knowledge. *Journal of Economic Dynamics & Control, 28*, 1557–1575.
Cowan, R., David P. A., & Foray, D. (2000). The explicit economics of knowledge: Codification and tacitness. *Industrial and Corporate Change, 9*, 211–253.
Dosi, G., & Marengo, L. (1994). Toward a theory of organizational competencies. In R. W. England (Ed.), *Evolutionary concepts in contemporary economics* (pp. 157–178). Ann Arbor, MI: Michigan University Press.
Durkheim, E. (1972). *Selected writings*. Cambridge, England: Cambridge University Press. (Original work published 1899).
Durkheim, E. (1997). *Suicide*. New York: Free Press. (Original work published in 1897)
Feldman, M. (1994). *The geography of innovation*. Boston, MA: Kluwer.
Feldman, M., & Florida, R. (1994). The geographic sources of innovation: Technological infrastructure and product innovation in the United States. *Annals of the Association of American Geographers, 84*, 210–229.
Feldman, S. P. (1997). The revolt against cultural authority: Power/knowledge as an assumption in organization theory. *Human Relations, 50*, 937–955.
Fleck, L. (1980). *Entstehung und Entwicklung einer wissenschaftlichen Tatsache. Einführung in die Lehre vom Denkstil und Denkkollektiv* [Emergence and development of a scientific fact: Introduction to the theory of styles of thinking style and collectives of thinkers]. Frankfurt am Main, Germany: Suhrkamp. (Original work published 1935)
Fletcher, J. (1849). Moral and educational statistics of England and Wales. *Journal of the Statistical Society of London, 12*, 151–176, 189–335.
Flyvbjerg, B. (1998). *Rationality and power: Democracy in practice*. Chicago, IL: University of Chicago Press.
Foote, K., Tóth, A., & Árvay, A. (2000). Hungary after 1989: Inscribing a new past on place. *The Geographical Review, 90*, 301–334.
Foucault, M. (1972). *The archaeology of knowledge*. New York: Pantheon.
Foucault, M. (1980). *Power/knowledge: Selected interviews and other writings, 1972–1977*. Brighton, England: Pantheon.
Franck, G. (1998). *Ökonomie der Aufmerksamkeit. Ein Entwurf* [The economics of attention: An outline]. Munich, Germany: Hanser.
Frantz, K. (1994). Washington Schools, Little White Man Schools und Indian Schools—bildungsgeographische Fragestellungen dargestellt am Beispiel der Navajo Indianerreservationen [Issues of educational geographics as exemplified by the Navajo Indian reservations]. *Die Erde, 125*, 299–314.
Freytag, T. (2003). *Bildungswesen, Bildungsverhalten und kulturelle Identität. Ursachen für das unterdurchschnittliche Ausbildungsniveau der hispanischen Bevölkerung in New Mexico* [Education, educational behavior, and cultural identity: Causes of the below-average level of education among the Hispanic population in New Mexico] (Heidelberger Geographische Arbeiten No. 118). Heidelberg, Germany: Selbstverlag des Geographischen Instituts.
Fujita, M., Krugman, P., & Venables, A. J. (1999). *The spatial economy: Cities, regions, and international trade*. Cambridge, MA: MIT.
Furet, F., & Ozouf, J. (1977). *Lire et Ecrire. L'Alphabétisation des Français de Calvin à Jules Ferry* [Reading and writing: Literacy of the French from Calvin to Jules Ferry]. 2 vols. Paris: Les Editions de Minuit.
Gadamer, H.-G. (1960). Wahrheit und Methode; Grundzüge einer philosophischen Hermeneutik. [Truth and methods: Principles of philosophical hermeneutics]. Tübingen, Germany: Mohr.
Gadamer, H.-G. (1987a). Rationalität im Wandel der Zeit [Rationality in the course of time]. In H.-G. Gadamer: *Gesammelte Werke* (Vol. 4, pp. 23–36). Tübingen, Germany: Mohr.

Gadamer, H.-G. (1987b). Theorie, Technik, Praxis [Theory, technologly, praxis]. In H.-G. Gadamer: *Gesammelte Werke* (Vol. 4, pp. 243–266), Tübingen, Germany: Mohr.

Galison, P. (1997). *Image and logic*. Chicago, IL: University of Chicago Press.

Galison, P. (1999). Buildings and the subject of science. In P. Galison & E. Thompson (Eds.), *The architecture of science* (pp. 1–25). Cambridge, MA: MIT.

Gamerith, W. (2006). Ethnizität und Bildungsverhalten. Ein kritisches Plädoyer für eine "Neue" Kulturgeographie [Ethnicity and educational behavior: A critical call for a "new" cultural geography]. In P. Meusburger & K. Kempter (Eds.), *Bildung und Wissensgesellschaft* (pp. 309–332). Heidelberger Jahrbücher No. 49. Heidelberg, Germany: Springer.

Gans, H. J. (2002). The sociology of space: A use-centered view. *City & Community, 1*, 325–335.

Gaspar, J., & Glaeser, E. L. (1998). Information technology and the future of cities. *Journal of Urban Economics, 43*, 136–156.

Geipel, R. (1965). *Sozialräumliche Strukturen des Bildungswesens. Studien zur Bildungsökonomie und zur Frage der gymnasialen Standorte in Hessen*. [Sociogeographical structures of the educational system: Studies on educational economics and the locations of secondary schools in Hessen]. Frankfurt am Main, Germany: Diesterweg.

Geipel, R. (1968). Der Standort der Geographie des Bildungswesens innerhalb der Sozialgeographie [The position of the geography of education within social geography]. K. Ruppert (Ed.), *Zum Standort der Sozialgeographie. Wolfgang Hartke zum 60. Geburtstag* (pp. 155–161). Münchener Studien zur Sozial- und Wirtschaftsgeographie, Vol. 4. Kallmünz, Germany: Lassleben.

Geipel, R. (1969). Bildungsplanung und Raumordnung als Aufgaben moderner Geographie [Educational planning and regional development as tasks of modern geography]. *Geographische Rundschau, 23*, 15–26.

Geipel, R. (1971). Die räumliche Differenzierung des Bildungsverhaltens. [Spatial disparities of educational attainment]. In *Bildungsplanung und Raumordnung* (pp. 47–61). Veröffentlichungen der Akademie für Raumforschung und Landesplanung, Forschungs- und Sitzungsberichte, No. 61. Hannover,Germany: Jaenecke.

Gertler, M. S. (2003). Tacit knowledge and the economic geography of context, or The undefinable tacitness of being (there). *Journal of Economic Geography, 3*, 75–99.

Geser, H. (1983). *Strukturformen und Funktionsleistungen sozialer Systeme* [Structural forms and functional performance of social systems]. Opladen, Germany: Westdeutscher Verlag.

Giddens, A. (1984a). *The constitution of society: Outline of the theory of structuralization*. Cambridge, England: Polity.

Giddens, A. (1984b). Space, time and politics in social theory. *Environment and Planning D: Society and Space, 2*, 123–132.

Gieryn T. F. (1999). *Cultural boundaries of science: Credibility on the line*. Chicago, IL: University of Chicago Press.

Gieryn, T. F. (2000). A space for place in sociology. *Annual Reviews of Sociology, 26*, 463–496.

Gieryn, T. F. (2001). Sociology of science. In N. J. Smelser & P. B. Baltes (Eds.), *International encyclopedia of the social & behavioral sciences* (Vol. 20, pp. 13692–13698). Amsterdam, The Netherlands: Elsevier.

Gieryn, T. F. (2002a). Give place a chance: Reply to Gans. *City & Community, 1*, 341–343.

Gieryn, T. F. (2002b). Three truth-spots. *Journal of the History of the Behavioral Sciences, 38*, 113–132.

Gieryn, T. F. (2002c). What buildings do. *Theory and Society, 31*, 35–74.

Gigerenzer, G. (2001). The adaptive toolbox. In G. Gigerenzer & R. Selten (Eds.), *Bounded rationality: The adaptive toolbox* (pp. 37–50). Cambridge, MA: MIT.

Gigerenzer, G., & Selten, R. (2001). Rethinking rationality. In G. Gigerenzer & R. Selten (Eds.), *Bounded rationality: The adaptive toolbox* (pp. 1–12). Cambridge, MA: MIT.

Gimbel, J. (1990). *Science, technology, and reparations: Exploitation and plunder in postwar Germany*. Stanford, CA: Stanford University Press.

Glückler, J., & Armbrüster, T. (2003). Bridging uncertainty in management consulting: The mechanisms of trust and networked reputation. *Organization Studies, 24*, 269–297.
Goddard, J. B. (1971). Office communications and office location: A review of current research. *Regional Studies, 5*, 263–280.
Goddard, J. B., & Morris, D. (1976). The communications factor in office decentralization. *Progress in Planning, 6*, 1–80.
Goddard, J. B., & Pye, R. (1977). Telecommunications and office location. *Regional Studies, 11*, 19–30.
Goss, J. (1988). The built environment and social theory. *Professional Geographer, 40*, 392–403.
Gottmann, J. (1980). Confronting centre and periphery. In J. Gottmann (Ed.), *Centre and periphery: Spatial variation in politics* (pp. 11–25). London: Sage.
Gregory, D. (1994). *Geographical imaginations*. Oxford, England: Basil Blackwell.
Gregory, D. (1998). Power, knowledge and geography. *Geographische Zeitschrift, 86*, 70–93.
Gregory, D. (2000). Cultures of travel and spatial formations of knowledge. *Erdkunde, 54*, 297–319.
Gregory, D. (2004). *The colonial present: Afghanistan, Palestine, Iraq*. Malden, MA: Blackwell.
Gruber, H. (2001). Acquisition of expertise. In N. J. Smelser & P. B. Baltes (Eds.), *International encyclopedia of the social & behavioral sciences* (Vol. 8, pp. 5145–5150). Amsterdam, The Netherlands: Elsevier.
Günzel, S. (2006). Einleitung [Introduction]. In J. Dünne & S. Günzel (Eds.), *Raumtheorie. Grundlagentexte aus Philosophie und Kulturwissenschaften* (pp. 19–43). Frankfurt am Main, Germany: Suhrkamp.
Habraken, N. J. (1998). *The structure of the ordinary*. Cambridge, MA: MIT.
Hacking, I. (1985). Styles of scientific reasoning. In J. Rajchman & C. West (Eds.), *Postanalytic philosophy* (pp. 145–165). New York: Columbia University Press.
Hacking, I. (2002). *Historical ontology*. Cambridge, MA: Harvard University Press.
Hägerstrand, T. (1966). Aspects of the spatial structure of social communication and the diffusion of information. *Papers and Proceedings of the Regional Science Association, 16*, 27–42.
Hägerstrand, T., & Kuklinski, A. R. (Eds.). (1971). Information systems for regional development: A seminar. General papers (Lund Studies in Geography, Series B., No. 37). Lund, Sweden: Gleerup.
Haraway, D. (1988). Situated knowledges: The science question in feminism and the privilege of partial perspective. In M. Biagioli (Ed.), *The science studies reader* (pp. 172–188). New York: Routledge.
Hard, G. (1999). Raumfragen [Questions about space]. In G. Kohlhepp, A. Leidlmair, & F. Scholz (Series Eds.) & P. Meusburger (Vol. Ed.), *Erdkundliches Wissen: Vol. 130. Handlungszentrierte Sozialgeographie. Benno Werlen's Entwurf in kritischer Diskussion* (pp. 133–162). Stuttgart, Germany: Steiner.
Hard, G. (2002). *Landschaft und Raum. Aufsätze zur Theorie der Geographie 1* [Landscape and space: Essays on the theory of geography 1]. Osnabrücker Studien zur Geographie 22. Osnabrück, Germany: Universitätsverlag Rasch.
Harouel, J. L. (2001). Sociology of culture. In N. J. Smelser & P. B. Baltes (Eds.), *International encyclopedia of the social & behavioral sciences* (Vol. 5, pp. 3179–3184). Amsterdam, The Netherlands: Elsevier.
Harvey, D. (1969). *Explanations in geography*. London: Edward Arnold.
Harvey, D. (1972). Revolutionary and counter-revolutionary theory in geography and the problem of ghetto formation. *Antipode, 4*, 1–13.
Harvey, D. (1973). *Social justice and the city*. London: Edward Arnold.
Harvey, D. (2005). *Space as a key word* (Hettner Lectures Vol. 8, pp. 93–115). Stuttgart, Gemany: Steiner.
Hasse, J. (1998). Zum Verhältnis von Raum und Körper in der Informationsgesellschaft [On the relation between space and body in the information society]. *Geographica Helvetica, 53*(2), 51–59.

Hatano G., & Oura, Y. (2001). Culture-rooted expertise: Psychological and educational aspects. In N. J. Smelser & P. B. Baltes (Eds.), *International encyclopedia of the social & behavioral sciences* (Vol. 5, pp. 3173–3176). Amsterdam, The Netherlands: Elsevier.

Hatano G., & Takahashi, K. (2001). Cultural diversity, human development, and education. In N. J. Smelser & P. B. Baltes (Eds.), *International encyclopedia of the social & behavioral sciences* (Vol. 5, pp. 3041–3045). Amsterdam, The Netherlands: Elsevier.

Hayden, D. (1995). *The power of place: Urban landscapes as public history.* Cambridge MA: MIT.

Hayden, D. (2001). Power of place. In N. J. Smelser & P. B. Baltes (Eds.), *International encyclopedia of the social & behavioral sciences* (Vol. 17, pp. 11451–11455). Amsterdam, The Netherlands: Elsevier.

Hayek, F. A. von (1937). Economics and knowledge. *Economica (New Series), 4*, 33–54.

Hayek, F. A. von (1945). The use of knowledge in society. *American Economic Review, 35*, 519–530.

Heffernan, M. (1988). The geography of ignorance in early-modern France. *Transactions, Institute of British Geographers (New Series), 13*, 228–231.

Heffernan, M. (1989). Literacy and geographical mobility in provincial France: Some evidence from the Départment of Ille-et-Vilaine. *Local Population Studies, 42*, 32–42.

Henkel, D., & Herkommer, B. (2004). Gemeinsamkeiten räumlicher und zeitlicher Strukturen und Veränderungen [Commonalities of spatial and temporal structures and changes]. In W. Siebel (Ed.), *Die europäische Stadt* (pp. 52–66). Frankfurt am Main, Germany: Suhrkamp.

Herskovits, M. J. (1948). Man and his works: The science of cultural anthropology. New York: A. Knopf.

Hetherington, K. (1997). In place of geometry: The materiality of place. In K. Hetherington & R. Munro (Eds.), *Ideas of difference: Social spaces and the labour of division* (Sociological Review Monograph Series, pp. 183–199). Oxford, England: Blackwell.

Hillier, B., & Hanson, J. (1984). *The social logic of space.* Cambridge, England: Cambridge University Press.

Hölscher, T. (2006). Macht, Raum und visuelle Wirkung: Auftritte römischer Kaiser in der Stadtarchitektur von Rom [Power, space, and visual effect: Presentations of Roman emperors in the urban architecture of Rome]. In J. Maran, C. Juwig, H. Schwengel, & U. Thaler (Eds.), *Constructing power: Architecture, ideology and social practice* (pp. 185–205). Hamburg, Germany: LIT.

Hohenstein, G. (1971). Manager und Management [Managers and management]. In *Management-Enzyklopädie, das Management Wissen unserer Zeit in 6 Bänden,Vol 4* (pp. 366–372). Munich, Germany: Verlag Moderne Industrie.

Hoyler, M. (1996). Anglikanische Heiratsregister als Quellen historisch-geographischer Alphabet isierungsforschung [Anglican marriage registries as sources in research on historiogeographic literacy]. In D. Barsch, W. Fricke, & P. Meusburger (Series & Vol. Eds.), *Heidelberger Geographische Arbeiten: Vol. 100. 100 Jahre Geographie an der Ruprecht-Karls-Universität (1895–1995)* (pp. 174–200). Heidelberg, Germany: Selbstverlag des Geographischen Instituts.

Hoyler, M., & Jöns, H. (2005). Themenorte vernetzt gedacht: Reflexionen über iconoclashes und den Umgang mit Repräsentationen in der Geographie [Thinking of thematic places as networked: Reflections on iconoclashes and ways of dealing with representations in geography]. In M. Flitner & J. Lossau (Eds.), *Themenorte* (pp. 183–200). (Geographie No. 17). Münster, Germany: LIT.

Hoyningen-Huene, P. (1987). Contexts of discovery and contexts of justification. *Studies in History and Philosophy of Science, 18*, 501–515.

Ibert, O. (2007). Towards a geography of knowledge creation: The ambivalence between 'knowledge as an object' and 'knowing in practice'. *Regional Studies, 41*, 103–114.

Jahnke, H. (2004). Welcher Raum für welches Wissen? Beobachtungen aus Berlin [Which space for which knowledge? Observations from Berlin]. *Berichte zur deutschen Landeskunde, 78*, 269–282.

Jewett, R., & Lawrence, J. S. (2003). *Captain America and the crusade against evil: The dilemma of zealous nationalism*. Grand Rapids, MI: W. B. Eerdmans.

Jones, K. (2001). Trust: Philosophical aspects. In N. J. Smelser & P. B. Baltes (Eds.), *International encyclopedia of the social & behavioral sciences* (Vol. 23, pp. 15917–15922). Amsterdam, The Netherlands: Elsevier.

Jöns, H. (2003). *Grenzüberschreitende Mobilität und Kooperation in den Wissenschaften. Deutschlandaufenthalte US-amerikanischer Humboldt-Forschungspreisträger aus einer erweiterten Akteursnetzwerkperspektive* [Cross-boundary mobility and cooperation in the sciences: U.S. Humboldt Research Award winners in Germany from an expanded actor-network perspective] (Heidelberger Geographische Arbeiten No. 116). Heidelberg, Germany: Selbstverlag des Geographischen Instituts.

Jöns, H. (2006). Dynamic hybrids and the geographies of technoscience: Discussing conceptual resources beyond the human/non-human binary. *Social & Cultural Geography, 7*, 559–580.

Jöns, H. (2007). Transnational mobility and the spaces of knowledge production: A comparison of different academic fields. *Social Geography Discussions, 3*, 79–119.

Juwig, C. (2006). Figuren des Imaginären: Imaginationen und Architektur in karolingischer Zeit [Figures of the imaginary: Imaginations and architecture in Carolingian times]. In J. Maran, C. Juwig, H. Schwengel, & U. Thaler (Eds.), *Constructing power: Architecture, ideology and social practice* (pp. 207–240). Hamburg, Germany: LIT.

Keeble, D., & Wilkinson F. (1999). Collective learning and knowledge development in the evolution of regional clusters of high technology SMEs in Europe. *Regional Studies, 33*, 295–303.

Keith, M., & Pile, S. (Eds.). (1993). *Place and the politics of identity*. London: Routledge.

King, D. (1997). *Stalins Retuschen. Foto- und Kunstmanipulationen in der Sowjetunion* [Stalin's retouchings: Manipulations of photographs and art in the Soviet Union]. Hamburg, Germany: Hamburger Edition.

Kline, S. J., & Rosenberg, N. (1986). An overview of innovation. In R. Landau & N. Rosenberg (Eds.), *The positive sum strategy: Harnessing technology for economic growth* (pp. 275–305). Washington, DC: National Academy Press.

Klüter, H. (1986). *Raum als Element sozialer Kommunikation* [Space as element of social communication]. (Giessener Geographische Schriften 60). Giessen, Germany: Selbstverlag des Geographischen Instituts.

Klüter, H. (1999). Raum und Organisation [Space and organisation]. In G. Kohlhepp, A. Leidlmair, & F. Scholz (Series Eds.) & P. Meusburger (Vol. Ed.), *Erdkundliches Wissen: Vol. 130. Handlungszentrierte Sozialgeographie. Benno Werlen's Entwurf in kritischer Diskussion* (pp. 187–212). Stuttgart, Germany: Steiner.

Klüter, H. (2003). Raum als Umgebung [Space as an environment]. In G. Kohlhepp, A. Leidlmair, & F. Scholz (Series Eds.) & P. Meusburger & T. Schwan (Vol. Eds.), *Erdkundliches Wissen: Vol. 135. Humanökologie. Ansätze zur Überwindung der Natur-Kultur-Dichotomie* (pp. 217–238). Stuttgart, Germany: Steiner.

Knoke, K. (1996). *Bold New York: The essential road map to the twenty-first century*. New York: Kodansha.

Knorr-Cetina, K. (1992). The couch, the cathedral, and the laboratory: On the relationship between experiment and laboratory in science. In A. Pickering (Ed.), *Science as practice and culture* (pp. 113–138). Chicago, IL: University of Chicago Press.

Knorr-Cetina, K. (1999). *Epistemic cultures: How the sciences make knowledge*. Cambridge, MA: Harvard University Press.

Knorr-Cetina, K. (2001). Postsocial relations: Theorizing sociality in a postsocial environment. In B. Smart & G. Ritzer (Eds.), *Handbook of social theory* (pp. 520–537). London: Sage.

Koch, A. (2003). Raumkonstruktionen [Constructions of space]. In G. Kohlhepp, A. Leidlmair, & F. Scholz (Series Eds.) & P. Meusburger & T. Schwan (Vol. Eds) *Erdkundliches Wissen, Vol. 135. Humanökologie. Ansätze zur Überwindung der Natur Kultur Dichotomie* (pp. 175–196). Stuttgart, Germany: Steiner.

Konrád G., & Szelényi, I. (1978). *Die Intelligenz auf dem Weg zur Klassenmacht* [Intelligence on the way to class power]. Frankfurt am Main, Germany: Suhrkamp.

Kramer C. (1993). *Die Entwicklung des Standortnetzes von Grundschulen im ländlichen Raum. Vorarlberg und Baden-Württemberg im Vergleich* [The development of the local network of primary schools in rural areas: Vorarlberg and Baden-Württemberg compared] (Heidelberger Geographische Arbeiten No. 93). Heidelberg, Germany: Selbstverlag des Geographischen Instituts.

Kröcher, U. (2007). *Die Renaissance des Regionalen. Zur Kritik der Regionalisierungseuphorie in Ökonomie und Gesellschaft.* [The renaissance of the regional: A critique of the regionalization euphoria in business and society] (Raumproduktionen: Theorie und gesellschaftlich Praxis No. 2). Münster, Germany: Westfälisches Dampfboot.

Kuhn, T. (1962). *The structure of scientific revolutions.* Chicago, IL: University of Chicago Press.

Lagendijk, A (2001). Three stories about regional salience: 'regional worlds', 'political mobilisation', and 'performativity'. *Zeitschrift für Wirtschaftsgeographie, 45*, 139–158.

Lam, A. (2000). Tacit knowledge, organizational learning and societal institutions: An integrated framework. *Organization Studies, 21*, 487–513.

Latour, B. (1987). *Science in action: How to follow scientists and engineers through society.* Cambridge, MA: Harvard University Press.

Latour, B. (1999). *Pandora's hope: Essays on the reality of science studies.* Cambridge, MA: Harvard University Press.

Latour, B. (2001). Eine Soziologie ohne Objekt. Anmerkungen zur Interobjektivität [A sociology without an object: Comments on interobjectivity]. *Berliner Journal für Soziologie, 11*, 237–252.

Latour, B. (2002). What is iconoclash? Or is there a world beyond the image wars? In B. Latour & P. Weibel (Eds.), *Iconoclash: Beyond the image wars in science, religion, and art* (pp. 14–41). Cambridge, MA: MIT.

Latour, B., & Wolgar, S. (1979). *Laboratory life: The construction of scientific facts.* Beverley Hills, England: Sage.

Lawrence J. S., & Jewett, R. (2002). *The myth of the American superhero.* Grand Rapids, MI: W. B. Eerdmans.

Lawson, C., & Lorenz, E. (1999). Collective learning, tacit knowledge and regional innovative capacity. *Regional Studies, 33*, 305–317.

Lefebvre, H. (1991). *The production of space.* Oxford, England: Basil Blackwell.

Ley, D. (1977). Social geography and the taken-for-granted world. *Transactions of the Institute of British Geographers (New Series), 2*, 498–512.

Linde, H. (1972). *Sachdominanz in Sozialstrukturen* [Dominance of expertise in social structures]. Tübingen, Germany: Mohr.

Lippuner, R. (2005). *Raum, Systeme, Praktiken. Zum Verhältnis von Alltag, Wissenschaft und Geographie* [Space, systems, practices: On the relation between everyday life, science, and geography] (Sozialgeographische Bibliothek No. 2). Stuttgart, Germany: Steiner.

Lippuner, H., & Lossau, J. (2005). In der Raumfalle. Eine Kritik des spatial turn in den Sozialwissenschaften [In the space trap: A critique of the spatial turn in the social sciences]. In G. Mein & M. Riegler-Ladich (Eds.), *Soziale Räume und kulturelle Praktiken* (pp. 47–64). Bielefeld, Germany: transcript-Verlag.

Livingstone, D. (1987). *Darwin's forgotten defenders: The encounter between evangelical theology and evolutionary thought.* Grand Rapids, MI: Eerdmans.

Livingstone, D. (1995). The spaces of knowledge: Contributions towards a historical geography of science. *Environment and Planning D: Society and Space, 13*, 5–34.

Livingstone, D. (2000). Making space for science. *Erdkunde, 54*, 285–296.

Livingstone, D. (2002). *Knowledge, space and the geographies of science* (Hettner Lectures Vol. 5, pp. 7–40). Heidelberg, Germany: Heidelberg University Department of Geography.

Livingstone, D. (2003). *Putting science in its place: Geographies of scientific knowledge.* Chicago, IL: University of Chicago Press.

Livingstone, D. (2005). Text, talk and testimony: Geographical reflections on scientific habits—An afterword. *British Journal for the History of Science, 38*, 93–100.

Lorenzen, M., & Maskell, P. (2004). The cluster as a nexus of knowledge. In P. Cooke & A. Piccaluga (Eds.), *Regional economies as knowledge laboratories* (pp. 77–92). Cheltenham, England: Edward Elgar.
Lo, V. (2003). Local codes and global networks: Knowledge access as a location factor in the financial industry. In V. Lo & E. W. Schamp, (Eds.), *Knowledge, learning, and regional development* (pp. 61–81). (Wirtschaftsgeographie 24). Münster, Germany: LIT.
Löw, M. (2001). *Raumsoziologie* [Sociology of space]. Frankfurt am Main, Germany: Suhrkamp.
Lundvall, B. Å. (1988). Innovation as an interactive process: From producer-user interaction to the national system of innovation. In G. Dosi, C. Freeman, C. Nelson, R. R. Silverberg, & L. L. G. Soete (Eds.), *Technical change and economic theory* (pp. 349–369). London: Pinter.
Lundvall, B. Å. (1997). Information technology in the learning economy: Challenges for development strategies. *Communications and Strategies, 28*, 177–192.
Machlup, F. (1962). *The production and distribution of knowledge in the United States*. Princeton, NJ: Princeton University Press.
Mager, C. (2007). *HipHop, Musik und die Artikulation von Geographie* [Hip-hop, music, and the articulation of geography] (Sozialgeographische Bibliothek No. 8). Stuttgart, Germany: Steiner.
Maillat, D., Quévit, M., & Sen, L. (1993). Réseaux d'innovation et milieux innovateurs: un pari pour le développement régional [Innovation networks and innovative environments: Betting on regional development]. Neuchâtel, Switzerland: Edes.
Malecki, E. J. (1980). Corporate organization of R and D and the location of technological activities. *Regional Studies, 14*, 219–234.
Malecki, E. J. (1997). *Technology and economic development: The dynamics of local, regional and national competitiveness* (2nd ed.). Harlow, England: Longman.
Malecki, E. J. (2000). Knowledge and regional competitiveness. *Erdkunde, 54*, 334–351.
Malmberg, A., & Maskell, P. (2002). The elusive concept of localization economies: Towards a knowledge-based theory of spatial clustering. *Environment and Planning A, 34*, 429–449.
Mann, M. (1986). *The sources of social power: Vol. 1. A history of power from the beginning to A.D. 1760*. Cambridge, England: Cambridge University Press.
Maran, J. (2006). Architecture, power and social practice: An introduction. In J. Maran, C. Juwig, H. Schwengel, & U. Thaler. (Eds.), *Constructing power: Architecture, ideology and social practice* (pp. 9–14). Hamburg, Germany: LIT.
Markus, T. A. (2006). Piranesi's paradox: To build is to create asymmetries of power. In J. Maran, C. Juwig, H. Schwengel, & U. Thaler. (Eds.), *Constructing power: Architecture, ideology and social practice* (pp. 321–335). Hamburg, Germany: LIT.
Maskell, P., & Malmberg, A. (1999). Localised learning and industrial competitiveness. *Cambridge Journal of Economics, 23*, 167–185.
Massey, D. (1985). New directions in space. In D. Gregory & J. Urry (Eds.), *Social relations and spatial structures* (pp. 9–19). New York: St. Martin's.
Massey, D. (1999a). *Imagining globalisation: Power geometries of time-space* (Hettner Lectures Vol. 2, pp. 9–23). Heidelberg, Germany: Heidelberg University Department of Geography.
Massey, D. (1999b). *Philosophy and politics of spatiality: Some considerations* (Hettner Lectures, Vol. 2, pp. 27–42). Heidelberg, Germany: Heidelberg University Department of Geography.
Massey, D. (2005). *For space*. London: Sage.
Matthiesen, U. (2004). Wissen in Stadtregionen. Forschungsresultate und Streitfragen. Orientierungswissen und Handlungsoptionen [Knowledge in urban regions: Research results and controversies]. In U. Matthiesen (Ed.), *Stadtregion und Wissen. Analysen und Plädoyers für eine wissensbasierte Stadtpolitik* (pp. 11–28). Opladen, Germany: Westdeutscher Verlag.
McLuhan, M. (1964). *Understanding media: The extensions of man*. New York: McGraw-Hill.
Merikle, P. M. (2000). Subliminal perception. In E. Kazdin (Ed.), *Encyclopedia of psychology* (Vol. 17, pp. 497–499). New York: Oxford University Press.

Merikle, P. M., & Daneman, M. (1996). Memory for unconsciously perceived events: Evidence from anesthetized patients. *Consciousness and Cognition, 5*, 525–541.
Merikle, P. M., & Daneman, M. (1998). Psychological investigations of unconscious perception. *Journal of Consciousness Studies, 5*(1), 5–18.
Meusburger, P. (1976). Entwicklung, Stellung und Aufgaben einer Geographie des Bildungswesens. Eine Zwischenbilanz [Development, position, and tasks of a geography of education]. *Mitteilungen der Österreichischen Geographischen Gesellschaft, 118*, 9–54.
Meusburger, P. (1978). Regionale Unterschiede des Ausbildungsniveaus der Arbeitsbevölkerung. Zur regionalen Konzentration der Arbeitsplätze für Höherqualifizierte [Regional disparities of educational attainment of the working population: The regional concentration of workplaces for highly skilled persons]. In *Regionale Bildungsplanung im Rahmen der Entwicklungsplanung. Zusammenhänge zwischen Bildungs- und Beschäftigungssystem* (pp. 49–73). Veröffentlichungen der Akademie für Raumforschung und Landesplanung, Forschungs- und Sitzungsberichte, No. 127. Hannover, Germany: Schroedel.
Meusburger, P. (1980). *Beiträge zur Geographie des Bildungs- und Qualifikationswesens. Regionale und soziale Unterschiede des Ausbildungsniveaus der österreichischen Bevölkerung* [Contributions to the geography of education and qualification: Regional and social differences of educational achievement of the Austrian population] (Innsbrucker Geographische Studien No. 7). Innsbruck, Austria: Innsbruck University Department of Geography.
Meusburger, P. (1990). Die regionale und soziale Herkunft der Heidelberger Professoren zwischen 1850 und 1932 [The regional and social origin of Heidelberg's Professors between 1850 and 1932]. In D. Barsch, W. Fricke, & P. Meusburger (Series Eds.) & P. Meusburger and J. Schmude (Vol. Eds.), *Heidelberger Geographische Arbeiten: Vol. 88. Bildungsgeographische Arbeiten über Baden-Württemberg* (pp. 187–239). Heidelberg, Germany: Selbstverlag des Geographischen Instituts.
Meusburger, P. (1996a). Educational achievement, language of instruction, and school system as key elements of minority research. In K. Frantz & R. A. Auder (Eds.), *Ethnic persistence and change in Europe and America* (Veröffentlichungen der Universität Innsbruck No. 213, pp. 187–222). Innsbruck, Austria: University of Innsbruck.
Meusburger, P. (1996b). Zur räumlichen Konzentration von "Wissen und Macht" im realen Sozialismus [On the spatial concentration of "knowledge and power" in socialism as actually practiced]. In D. Barsch, W. Fricke, & P. Meusburger (Series and Vol. Eds), *Heidelberger Geographische Arbeiten: Vol. 100. 100 Jahre Geographie an der Ruprecht-Karls-Universität (1895–1995)* (pp. 216–236). Heidelberg: Selbstverlag des Geographischen Instituts.
Meusburger, P. (1998). *Bildungsgeographie. Wissen und Ausbildung in der räumlichen Dimension* [Geography of education: Knowledge and education in the spatial dimension]. Heidelberg, Germany: Spektrum Akademischer Verlag.
Meusburger, P. (1999). Subjekt—Organisation—Region. Fragen an die subjektzentrierte Handlungstheorie [Subject—organization—region: Questions for subject-centered action theory]. In G. Kohlhepp, A. Leidlmair, & F. Scholz (Series Eds.) & P. Meusburger (Vol. Ed.), *Erdkundliches Wissen: Vol. 130. Handlungszentrierte Sozialgeographie. Benno Werlens Entwurf in kritischer Diskussion* (pp. 95–132). Stuttgart, Germany: Steiner.
Meusburger, P. (2000). The spatial concentration of knowledge: Some theoretical considerations. *Erdkunde, 54*, 352–364.
Meusburger, P. (2001a). Geography of knowledge, education, and skills. In N. J. Smelser & P. B. Baltes (Eds.), *International encyclopedia of the social & behavioral sciences* (Vol. 12, pp. 8120–8126). Amsterdam, The Netherlands: Elsevier.
Meusburger, P. (2001b). The role of knowledge in the socio-economic transformation of Hungary in the 1990s. In P. Meusburger & H. Jöns (Eds.), *Transformations in Hungary. Essays in economy and society* (pp. 1–38). Heidelberg, Germany: Physica-Verlag.
Meusburger, P. (2003). Culture, pouvoir et scolarité dans les États-nations multi-ethniques. Contextes culturels et géographie du savoir [Culture, power, and educational attainment in the multiethnic nation-states: Cultural contexts and geography of knowledge]. *Géographie et Cultures, 47*, 67–84.

Meusburger, P. (2005). Sachwissen und symbolisches Wissen als Machtinstrument und Konfliktfeld. Zur Bedeutung von Worten, Bildern und Orten bei der Manipulation des Wissens [Factual and symbolic knowledge as an instrument of power and a field of conflict: On the meaning of words, images, and places in the manipulation of knowledge]. *Geographische Zeitschrift, 93*, 148–164.

Meusburger, P. (2007). Macht, Wissen und die Persistenz von räumlichen Disparitäten [Power, knowledge, and the persistence of spatial disparities]. In I. Kretschmer (Ed.), *Das Jubiläum der Österreichischen Geographischen Gesellschaft. 150 Jahre (1856–2006)* (pp. 99–124). Vienna, Austria: Österreichische Geographische Gesellschaft.

Mintzberg, H. (1979). *The structuring of organizations: A synthesis of the research*. Englewood Cliffs. NJ: Prentice Hall.

Miggelbrink, J. (2002). Konstruktivismus? "Use with caution" ... Zum Raum als Medium der Konstruktion gesellschaftlicher Wirklichkeit [Constructivism? "Use with caution" ... On space as a medium in constructing social reality]. *Erdkunde, 56*, 337–350.

Morgan, K. (1997). The learning region: Institutions, innovation and regional renewal. *Regional Studies, 31*, 491–503.

Naisbitt, R. (1995). *The global paradox*. New York: Avon.

Naylor, S. (2002). The field, the museum and the lecture hall: The spaces of natural history in Victorian Cornwall. *Transactions of the Institute of British Geographers (New Series), 27*, 494–513.

Naylor, S. (2005a). Historical geography: Knowledge, in place and on the move. *Progress in Human Geography, 29*, 626–634.

Naylor, S. (2005b). Introduction: Historical geographies of science—Places, contexts, cartographies. *British Journal for the History of Science, 38*, 1–12.

Negroponte, N. (1995). *Being digital*. New York: Vintage.

Nelson, R. R. (1959a). The economics of invention: A survey of the literature. *Journal of Business, 32*, 101–127.

Nelson, R. R. (1959b). The simple economics of basic scientific research. *Journal of Political Economy, 67*, 297–306.

Nelson, R. R. (1962). The link between science and invention: The case of the transistor. In National Bureau of Economic Research (Ed.), *The rate and direction of inventive activity: Economic and social factors* (pp. 549–583). Princeton, NJ: Princeton University Press.

Nelson, R. R., & Winter, S. G. (1977). In search of a useful theory of innovation. *Research Policy, 6*, 36–76.

Nietzsche, F. (1888). *Der Wille zur Macht* [The will to power] (2nd ed.). Leipzig, Germany: Naumann.

Nonaka, I. (1994). A dynamic theory of organizational knowledge creation. *Organization Science, 5*, 14–37.

Nonaka, I., & Takeuchi, N. (1995). *The knowledge-creating company*. Oxford, England: Oxford University Press.

Nonaka, I., Toyama, R., & Nagata, A. (2000). A firm as a knowledge-creating entity: A new perspective on the theory of the firm. *Industrial and Corporate Change, 9*, 1–20.

Noteboom, B. (2000). *Learning and innovation in organizations and economies*. Oxford, England: Oxford University Press.

Numbers, R. L., & Stenhouse, J. (Eds.). (2001). *Disseminating Darwinism: The role of place, race, religion, and gender*. Cambridge, England: Cambridge University Press.

Ophir, A., & Shapin, S. (1991). The place of knowledge: The spatial setting and its relations to the production of knowledge [Special issue]. *Science in Context, 4*, 3–21.

Panayides, A., & Kern C. R. (2005). Information technology and the future of cities: An alternative analysis. *Urban Studies, 42*, 163–167.

Pavitt, K. (1998). Technologies product and organisation in the innovative firm: What Adam Smith tells us and Joseph Schumpeter does not. *Industrial and Corporate Change, 7*, 433–452.

Perrig, W., Wippich, W., & Perrig-Chiello, P. (1993). Unbewusste Informationsverarbeitung [Unconscious information processing]. Bern, Switzerland: Huber.
Pickles, J. (1985). *Phenomenology, science and geography: Spatiality and the human sciences*. Cambridge, England: Cambridge University Press.
Pickering, A. (1995). *The mangle of practice: Time agency and science*. Chicago, IL: Chicago University Press.
Piccolomini, A. P. (1965). *Caeremoniale romanum of Agostino Patrizi, Piccolomini: The first edition, Venice 1516, to which is appended Patrizi's original preface, rediscovered by Jean Mabillon and published by him in the second volume of his museum Italicum, Lutetiae Parisiorum 1689*. Ridgewood, NJ: Gregg. (Original work published 1516, Rituum ecclesiasticorum sive sacrarum cerimoniarium s. s. romanae ecclesiae: Libri tres non ante impressi, edited by C. Marcello, Venice, Italy: Leonardo Laureano)
Polanyi, M. (1958). Personal knowledge: towards a post-critical philosophy. London: Routledge & Kegan Paul.
Polanyi, M. (1967). *The tacit dimension*. New York: Doubleday.
Polanyi, M. (1985). *Implizites Wissen* [Implicit knowledge]. Frankfurt am Main, Germany: Suhrkamp.
Popper, K. (1934). *Logik der Forschung* [The Logic of research]. Tübingen, Germany: Mohr.
Porter, M. E. (2001). Regions and the new economics of competition. In A. J. Scott (Ed.), *Global city-regions: Trends, theory, policy* (pp. 139–157). Oxford, England: Oxford University Press.
Powell, R. (2007). Geographies of science: Histories, localities, practices, futures. *Progress in Human Geography, 31*, 309–329.
Pratt, M. L. (1992). *Imperial eyes: Travel writing and transculturation*. London: Routledge.
Pred, A. (1973). *Urban growth and the circulation of information*. Cambridge, MA: Harvard University Press.
Pred, A. (1984). Place as historically contingent process: Structuration and the time-geography of becoming places. *Annals of the Association of American Geographers, 74*, 279–297.
Pucci, M. (2006). Enclosing open spaces: The organisation of external areas in Syro-Hittite architecture. In J. Maran, C. Juwig, H. Schwengel, & U. Thaler (Eds.), *Constructing power: Architecture, ideology and social practice* (pp. 169–184). Hamburg, Germany: LIT.
Radner, R. (1987). Uncertainty and general equilibrium. In J. Eatwell, M. Milgrate, & P. Newman (Eds.), *The new Palgrave: A dictionary of economics* (Vol. 4, pp. 734–741). London: Macmillan.
Rapoport, A. (1982). *The meaning of the built environment: A nonverbal communication approach*. Beverly Hills, CA: Sage.
Reber, A. S. (1993). *Implicit learning and tacit knowledge: An essay on the cognitive unconscious*. (Oxford Psychology Series 19). New York: Clarendon.
Relph, E. (1976). *Place and placelessness*. London: Pion.
Renn, O. (2001). Science and technology studies: Experts and expertise. In N. J. Smelser & P. B. Baltes (Eds.), *International encyclopedia of the social & behavioral sciences* (Vol. 20, pp. 13647–13654), Amsterdam, The Netherlands: Elsevier.
Rupke, N. A. (2005). *Alexander von Humboldt, a metabiography*. Frankfurt am Main, Germany: Lang.
Sack, R. D. (1980). *Conceptions of space in social thought: A geographic perspective*. London: Macmillan.
Sahr W. D. (2006). Religion und Szientismus in Brasilien. Versuch (Essay) über eine dekonstruktivistische Regionalgeographie des Wissens [Religion and scientism in Brazil: Essay on a deconstructivist regional geography of knowledge]. *Geographische Zeitschrift, 94*, 27–42.
Schaffer, S. (1991). The eighteenth Brumaire of Bruno Latour. *Studies in History and Philosophy of Science, 22*, 174–192.
Schaffer, S. (1998). Physics laboratories and the Victorian country house. In C. Smith & J. Agar (Eds.), *Making space for science: Territorial themes in the shaping of knowledge* (pp. 149–180). Basingstoke, England: Macmillan.

Schamp, E. W., & Lo V. (2003). Knowledge, learning, and regional development: An introduction. In V. Lo & E. W. Schamp (Eds.), *Knowledge, learning, and regional development* (pp. 1–12). (Wirtschaftsgeographie 24). Münster, Germany: LIT.

Schatzki, T. (2007). *Martin Heidegger: Theorist of space* (Sozialgeographische Bibliothek No. 6). Stuttgart, Germany: Steiner.

Scheler, M. (1926). *Die Wissensformen und die Gesellschaft* [The forms of knowledge and society]. Leipzig, Germany: Der Neue-Geist Verlag.

Schlögel, K. (2003). *Im Raume lesen wir die Zeit: über Zivilisationsgeschichte und Geopolitik* [We read time in spaces: On the history of civilization and geopolitics]. Munich, Germany: Carl Hanser.

Schmid, C. (2005). *Stadt, Raum und Gesellschaft. Henri Lefebvre und die Theorie der Produktion des Raumes* [City, space, and society: Henri Lefebvre and the theory of the production of space] (Sozialgeographische Bibliothek No. 1). Stuttgart, Germany: Steiner.

Schmude, J. (1988). *Die Feminisierung des Lehrberufs an öffentlichen, allgemeinbildenden Schulen in Baden-Württemberg, eine raum-zeitliche Analyse* [The femininization of the teaching profession at public schools providing general education in Baden-Württemberg: A time-space analysis] (Heidelberger Geographische Arbeiten No. 87). Heidelberg, Germany: Selbstverlag des Geographischen Instituts.

Schultz, T. W. (1960). Capital formation by education. *Journal of Political Economy, 68*, 571–583.

Schultz, T. W. (1963). *Education and economic growth*. New York: Columbia University Press.

Schumpeter, J. (1912). *Theorie der wirtschaftlichen Entwicklung*. Leipzig, Germany: Dunkker und Humblot.

Schwan, T. (2003). Clash of imaginations—Erfahrungswissenschaftliches Menschenbild versus postmoderne Konstruktionen [Clash of imaginations—Empirical view of human versus postmodern constructions]. In G. Kohlhepp, A. Leidlmair, & F. Scholz (Series Eds.) & P. Meusburger & T. Schwan (Vol. Eds.), *Erdkundliches Wissen: Vol. 135. Humanökologie. Ansätze zur Überwindung der Natur-Kultur-Dichotomie* (pp. 161–173). Stuttgart, Germany: Steiner.

Shapin, S. (1988). The house of experiment in seventeenth-century England. *Isis, 79*, 373–404.

Shapin, S. (1991). 'The mind is its own place': Science and solitude in seventeenth-century England. *Science in Context, 4*, 191–218.

Shapin, S. (1998). Placing the view from nowhere: Historical and sociology problems in the location of science. *Transactions of the Institute of British Geographers (New Series), 23*, 5–12.

Shapin, S. (2001). Truth and credibility: Science and the social study of science. In N. J. Smelser & P. B. Baltes (Eds.), *International encyclopedia of the social & behavioral sciences* (Vol. 23, pp. 15926–15932). Amsterdam, The Netherlands: Elsevier.

Shapin, S., & Schaffer, S. (1985). *Leviathan and the airpunp: Hobbes, Boyle, and the experimental life*. Princeton, NJ: Princeton University Press.

Simon, H. A. (1956). Rational choice and the structure of environments. *Psychological Review, 63*, 129–138.

Simon, H. A. (1962). The architecture of complexity. *Proceedings of the American Philosophical Society, 106*, 467–482.

Singh, A. (1994). Global economic change: Skills and international competitiveness. *International Labour Review, 133*, 167–183.

Soja, E. (1980). The socio-spatial dialectics. *Annals of the Association of American Geographers, 70*, 207–225.

Soja, E. (1985). The spatiality of social life: Towards a transformative retheorisation. In D. Gregory & J. Urry (Eds.), *Social relations and spatial structures* (pp. 90–127). London: Macmillan.

Soja, E. (2003). *Postmodern geographies* (8th ed.). London: Verso.

Spillman, L. (1998). When do collective memories last? *Social Science History, 22*, 445–477.

Spinner, H. F. (1994). *Die Wissensordnung: Ein Leitkonzept für die dritte Grundordnung des Informationszeitalters* [Studies on the knowledge order: Vol. 1. A concept to guide the third

fundamental order of the information age] (Studien zur Wissensordnung No. 1). Opladen, Germany: Leske & Budrich.
Squire, L. R. (1986). Mechanisms of memory. *Science, 232*, 1612–1619.
Stam, E., & Wever E. (2003). Propinquity without community. Spatial transfer of knowledge in the Netherlands: A national and international comparison of collective learning in high-tech manufacturing and services. In V. Lo & E. W. Schamp (Eds.), *Knowledge, learning, and regional development* (pp. 38–60). (Wirtschaftsgeographie 24). Münster, Germany: LIT.
Stehr, N. (1994a). *Arbeit, Eigentum und Wissen. Zur Theorie von Wissensgesellschaften* [Work, property, and knowledge: On the theory of knowledge societies]. Frankfurt am Main, Germany: Suhrkamp.
Stehr, N. (1994b). *Knowledge societies*. London: Sage.
Stehr, N., & Meja, V. (Eds.). (2005). *Society and knowledge: Contemporary perspectives in the sociology of knowledge and science* (2nd ed.). New Brunswick, NJ: Transaction Books.
Stenhouse, J. (2001). Darwinism in New Zealand, 1859–1900. In R. L. Numbers & J. Stenhouse (Eds.), *Disseminating Darwinism: The role of place, race, religion, and gender* (pp. 61–90). Cambridge, England: Cambridge University Press.
Sternberg, R. (2007). Innovative new firms, embeddedness and regional development. *International Journal Entrepreneurship and Innovation Management, 7*, 445–461.
Storper, M. (1995). The resurgence of regional economics, ten years later: The region as a nexus of untraded interdependencies. *European Urban and Regional Studies, 2*, 191–221.
Storper, M., & Venables, A. J. (2004). *Buzz: Face-to-face contacts and the urban economy* (Hettner Lectures Vol. 7, pp. 43–66). Heidelberg, Germany: Heidelberg University Department of Geography.
Strassoldo, R. (1980). Centre-periphery and system boundary: Cultural perspectives. In J. Gottmann (Ed.), *Centre and periphery: Spatial variation in politics* (pp. 27–61). London: Sage.
Sztompka, P. (1999). *Trust: A sociological theory*. Cambridge, England: Cambridge University Press.
Sztompka, P. (2001). Trust: Cultural concerns. In N. J. Smelser & P. B. Baltes (Eds.), *International encyclopedia of the social & behavioral sciences* (Vol. 23, pp. 15913–15917). Amsterdam, The Netherlands: Elsevier.
Thorngren, B. (1970). How do contact systems affect regional development? *Environment and Planning, 2*, 409–427.
Thrift, N. J. (1983). On the determination of social action in space and time. *Environment and Planning D: Society and Space, 1*, 23–57.
Thrift, N. J. (1985). Flies and germs: a geography of knowledge. In D. Gregory & J. Urry (Eds.), *Social relations and spatial structures* (pp. 366–403). London: Macmillan.
Thrift, N. J. (1999). Steps to an ecology of place. In D. Massey, J. Allen, & P. Sarre (Eds.), *Human geography today* (pp. 295–322). Cambridge, England: Polity.
Thrift, N., Driver, F., & Livingstone D. (1995). The geography of truth. *Environment and Planning D: Society and Space, 13*, 1–3.
Tomiak, J. (Ed.). (1991). *Schooling, educational policy and ethnic identity: Comparative studies on governments and non-dominant ethnic groups in Europe, 1850–1940*. Dartmouth, England: Aldershot.
Törnqvist, G. (1968). *Flows of information and the location of economic activities* (Lund Studies in Geography, Series B, No. 30). Lund, Sweden: Gleerup.
Törnqvist, G. (1970). *Contact systems and regional development* (Lund Studies in Geography, Series B, No. 35). Lund, Sweden: Gleerup.
Toffler, A. (1980). *The third wave*. New York: Bantam.
Townley, B. (1993). Foucault, power/knowledge, and its relevance for human resource management. *Academy of Management Review, 18*, 518–545.
Trueba, H. T. (1989). *Raising silent voices: Educating the linguistic minorities for the 21st century*. New York: Newbury House.
Tuan, Y. F. (1977). *Space and place: The perspective of experience*. Chicago, IL: University of Chicago Press.

Wassmann, J. (1998). Balinese spatial orientation: Some evidence for moderate linguistic relativity. *Journal of the Royal Anthropological Society (Man), 4*, 689–711.

Wassmann, J. (2003). Kognitive Ethnologie [Cognitive ethnology]. In B. Beer & H. Fischer (Eds.), *Ethnologie: Einführung und Überblick* (pp. 323–340). Berlin: Reimer.

Webber, M. M. (1964). The urban place and the nonplace urban realm. In M. M. Webber, J. W. Dyckman, D. L. Foley, A. Z. Guttenberg, W. L. C. Wheaton, & D. Bauer Wuster (Eds.), *Explorations into urban structure* (pp. 79–153). Philadelphia, PA: University of Phildelphia Press.

Webber, M. M. (1973). *Societal contexts of transportation and communication*. Berkeley, CA: Institute of Urban & Regional Development, University of California.

Weber, M. (1978). *Economy and society: An outline of interpretive sociology*. 2 Vols. (G. Roth & C. Wittich, Eds; E. Fischoff, H. Gerth, A. M. Henderson, F. Kolegar, C. Wright Mills, T. Parsons, M. Rheinstein, G. Roth, E. Shils, & C. Wittich, Trans.). Berkeley, CA: University of California Press. (Original work published 1922).

Weddigen T. (2006). Liturgical space as a social product: Anthropological aspects of the early modern Maiestas Pontificia in the sistine chapel. In J. Maran, C. Juwig, H. Schwengel, & U. Thaler (Eds.), *Constructing power: Architecture, ideology and social practice* (pp. 265–284). Hamburg, Germany: LIT.

Weichhart, P. (1996). Die Region—Chimäre, Artefakt oder Strukturprinzip sozialer Systeme? [The region—Chimera, artifact, or structural principle of social systems]. In G. Brunn (Ed.), *Region und Regionsbildung in Europa: Konzeptionen der Forschung und empirische Befunde* (pp. 25–43). (Schriftenreihe des Instituts für Europäische Regionalforschung, Vol. 1). Baden-Baden, Germany: Nomos.

Weichhart, P. (1999). Die Räume zwischen den Welten und die Welt der Räume. Zur Konzeption eines Schlüsselbegriffs der Geographie [The spaces between the worlds and the world of spaces: Toward inception of a key concept of geography]. In G. Kohlhepp, A. Leidlmair, & F. Scholz (Series Eds.) & P. Meusburger (Vol. Ed.), *Erdkundliches Wissen: Vol. 130. Handlungszentrierte Sozialgeographie: Benno Werlens Entwurf in kritischer Diskussion* (pp. 67–94). Stuttgart, Germany: Steiner.

Weichhart, P. (2003). Gesellschaftlicher Metabolismus und Action Settings. Die Verknüpfung von Sach- und Sozialstrukturen im alltagsweltlichen Handeln [Social metabolism and action settings: The link between technical and social structures in everyday action]. In In G. Kohlhepp, A. Leidlmair, & F. Scholz (Series Eds.) & P. Meusburger & T. Schwan (Vol. Eds.), *Erdkundliches Wissen: Vol. 135. Humanökologie: Ansätze zur Überwindung der Natur-Kultur-Dichotomie* (pp. 15–44). Stuttgart, Germany: Steiner.

Weick, C. (1995). *Räumliche Mobilität und Karriere: Eine individualstatistische Analyse der baden-württembergischen Universitätsprofessoren unter besonderer Berücksichtigung demographischer Strukturen* [Spatial mobility and career: A statistical analysis of Baden-Württemberg university professors, with special focus on demographic structures] (Heidelberger Geographische Arbeiten No. 101). Heidelberg, Germany: Selbstverlag des Geographischen Instituts.

Weinberg, A. K. (1935). *Manifest destiny: A study of nationalist expansionism in American history*. Baltimore, MD: Johns Hopkins.

Wenger, E. (1998). *Communities of practice: Learning, meaning, and identity*. Cambridge, England: Cambridge University Press.

Werlen, B. (1987). *Gesellschaft, Handlung und Raum. Grundlagen handlungstheoretischer Sozialgeographie* [Society, Action, and space: Principles of social geography based on action theory] (Erdkundliches Wissen No. 89). Stuttgart, Germany: Steiner.

Werlen, B. (1993). *Society, action and space: An alternative human geography*. London: Routledge.

Werlen, B. (1995). *Sozialgeographie alltäglicher Regionalisierungen. Band 1: Zur Ontologie von Gesellschaft und Raum* [Social geography of everyday regionalizations: Vol. 1. On the ontology of society and space] (Erdkundliches Wissen No. 116). Stuttgart, Germany: Steiner.

Werlen, B. (1997a). Social geography. In L. Embree, J. Kokelmans, & R. Zaner (Eds.), *The encyclopedia of phenomenology* (pp. 645–650). Dordrecht, The Netherlands: Kluwer.

Werlen, B. (1997b). *Sozialgeographie alltäglicher Regionalisierungen. Band 2: Globalisierung, Region und Regionalisierung* [Social geography of everyday regionalizations: Vol. 2. Globalization, region, and regionalization] (Erdkundliches Wissen No. 119). Stuttgart, Germany: Steiner.

Werlen, B. (1997c). *Gesellschaft, Handlung und Raum. Grundlagen handlungstheoretischer Sozialgeographie* [Society, action and space. Basics of an action-theoretical social geography]. 3. Ed., Stuttgart, Germany: Steiner Verlag.

Werlen, B. (1999). Political regionalism. In L. Embree (Ed.), *Alfred Schutz's theory of social sciences* (pp. 1–22). Dordrecht, The Netherlands: Kluwer.

Werlen, B. (2004a). "Region" and the process of regionalisation. In H. van Houtum, O. Kramsch, & W. Zierhofer (Eds.), *Bordering space* (pp. 47–60). London: Ashgate Publishers.

Werlen, B. (2004b). *Sozialgeographie. Eine Einführung* [Social geography: An introduction] (2nd ed.). Bern, Switzerland: UTB/Paul Haupt.

Werlen, B. (2008 in print). Körper, Raum und mediale Repräsentation [Body, space, and medial representation.]. In J. Döring & T. Thielmann (Eds.), *Spatial Turn. Das Raumparadigma in den Kultur- und Sozialwissenschaften*. Bielefeld, Germany: transcript-Verlag.

Westaway, E. J. (1974). The spatial hierarchy of business organizations and its implications for the British urban system. *Regional Studies, 8*, 145–155.

Williams, R. (1973). Base and superstructure in Marxist cultural theory. *New Left Review, 82*, 3–16.

Withers, C. W. J. (2002). The geography of scientific knowledge. In N. A. Rupke (Ed.), *Göttingen and the development of the natural sciences* (pp. 9–18). Göttingen, Germany: Wallstein.

Withers, C. W. (2004). Memory and the history of geographical knowledge: The commemoration of Mungo Park, African explorer. *Journal of Historical Geography, 30*, 316–339.

Wright, J. C. (2006). The social production of space and the architectural reproduction of society in the bronze age Aegean during the 2nd Millenium B.C.E. In J. Maran, C. Juwig, H. Schwengel, & U. Thaler (Eds.), *Constructing power: Architecture, ideology and social practice* (pp. 49–74). Hamburg, Germany:LIT.

Wunder, E. (2005). *Religion in der postkonfessionellen Gesellschaft. Ein Beitrag zur sozialwissenschaftlichen Theorieentwicklung in der Religionsgeographie* [Religion in the postconfessional society: On the development of social science theory in the geography of religion] (Sozialgeographische Bibliothek No. 5). Stuttgart, Germany: Steiner.

Wunder, E. (2006). Religionen—Schmieröl im Kampf der Kulturen? [Religions: Lubrication in the clash of cultures?]. *Geographie und Schule, 28*, 15–19.

Zare, R. N. (1997). Knowledge and distributed intelligence. *Science, 275*, 1047.

Zerubavel, E. (1997). *Social mindscapes: An invitation to cognitive sociology*. Cambridge, MA: Harvard University Press.

Chapter 3
Cultural Boundaries: Settled and Unsettled

Thomas F. Gieryn

Realms of knowledge meet at the boundaries—cultural boundaries. Sometimes they clash, and at other times they do not. When scientific knowledge bumps up against religion, or against politics, ideology, market logics, common sense, or poetry, the result may be explicit and often passionate debate over the exact location of the boundary and the implications of drawing the line here or there for issues of power, authority, allocations of resources, and truth. But not always. Sometimes the cultural boundaries that separate realms of knowledge sit there peaceably, with little manifest attention from anybody, structuring everyday practices without noticeable contestation or doubt. Whether cultural boundaries become the occasion for clash or for reconciled juxtaposition depends on where one chooses to look. That is, different kinds of *places*—physical sites, with bounded geographic location and distinctive recognizable physical form—either open cultural boundaries to contestation or prevent such an overt clash from happening.

Swidler's (1986) distinction between settled and unsettled historical periods can usefully be applied to cultural boundaries, such as those between science and religion. Settled boundaries are stable and secure, institutionalized and routinized, structuring and enabling as though on autopilot, needing little or no manifest attention from the people who live inside them with little hesitation or scrutiny. Unsettled boundaries move into the foreground of discursive consciousness. Their location and even their existence become a matter for people to negotiate explicitly as they reflect on the potentially wide-ranging implications of a boundary becoming real here or there. Settled boundaries, by contrast, have that reality. They exist in a tacit but durable and imposing state, and they shape behavior, interpretive understandings, and allocations of valued resources. Unsettled boundaries are up for grabs, the focus of dispute and contestation among social actors each trying to arrange cultural territories and landmarks into a map that best suits their interests and purposes. Only in an unsettled state does the intersection of realms of knowledge result in a clash over their boundaries. The invisibility of settled cultural boundaries precludes manifest consideration and argument.

The potency of scientific knowledge—its assertion of objective truth, its promise of progress, its image of political and moral neutrality—has incessantly brought it into contact with other spaces in the culturescape (Gieryn, 1999). Whether or not

P. Meusburger et al. (eds.), *Clashes of Knowledge*.

that contact is marked by clash or quiet coexistence depends, in part, on the physical places where the encounters between cultural spaces are reified. One nonobvious place of science in the United States is the Federal Building and Courthouse in Harrisburg, Pennsylvania (see Fig. 3.1). It was the site of the 2005 trial known as Kitzmiller v. Dover Area School District, where Judge John E. Jones III ruled that intelligent design (like creation science more generally) is religion and not real science. The court also found that members of the Dover School Board hid their religious intentions as they sought to incorporate theories of intelligent design in the science curriculum of their public schools, in violation (the judge ruled) of constitutional separations of church and state (the Establishment Clause). A more obvious place of science is the James H. Clark Center at Stanford University (see Fig. 3.2), home to the University's Bio-X Initiative, a prize-winning building designed by Norman Foster and named after the founder of Netscape. It opened in 2003, bringing together 40 to 50 faculty scientists from medicine, the life sciences, engineering, computer science, and physics to work in gleaming new labs and offices on problems of bioinformatics and new medical therapies.

The Harrisburg Federal Courthouse is indisputably the setting for a clash of knowledge: science versus religion, the next round (indeed, the trial was sometimes referred to as Scopes II). This place put the limelight on the cultural boundary between science and religion, repeatedly erasing and redrawing it as adversaries sought to use the force of law to secure legitimacy for boundaries that served their interests best. By contrast, Stanford's Clark Center renders the cultural boundaries

Fig. 3.1 The U.S. Federal Building and Courthouse in Harrisburg, Pennsylvania

Fig. 3.2 The James H. Clark Center, Stanford University

of science in a geographical and architectural form that suppresses the possibility of clash. It is a setting for watching science as its boundaries get settled both in and through the building itself, without dispute or apparent stakes. Scientists go about their daily research without giving much explicit thought to how the design and location of this building materializes and stabilizes cultural boundaries between science and various other realms of knowledge, sets of practices, and institutions.

Exactly what happened inside Judge Jones's courtroom in Harrisburg in fall 2005? Simply put, a clash of knowledge took the form of "boundary-work," that is, "discursive attributions of selected qualities to scientists, scientific methods and scientific claims for the purpose of drawing a rhetorical boundary between 'science' and some less authoritative residual non-science" (Gieryn, 1983, p. 782). Boundary-work consists of strategic and practical demarcations of science carried out by scientists, would-be scientists, journalists, judges, and ordinary folk. It is pursued not just by philosophers of science like Karl Popper (whose famous demarcation criteria are deployed often in boundary-work, as rivals exploit Popper's reputation to justify their rhetorical games of inclusion and exclusion). Boundary-work is triggered by contested credibility, where adversaries use cartographic depictions of cultural differences to legitimate their claims to authority (over knowledge of human origins) and control (over the contents of what gets taught in school science classes). In these discursive contests, advocates on each side construct a space for science by selectively attributing qualities and potentials to "science" in a manner that makes them appear to be squarely inside.

For example, Eric Rothschild, attorney for the plaintiffs, stated the following in his opening remarks on Day One of the Dover trial:

> There is no data or laboratory work demonstrating intelligent design. It is not a testable hypothesis. It misrepresents established scientific knowledge. Let's be perfectly clear: there is no controversy in the scientific community about the soundness of evolution and that intelligent design is not a scientific topic at all. (*Kitzmiller v. Dover Area School District*, 9/26/2005)

Later, Rothschild added:

> Science does not consider supernatural explanations because it has no way of observing, measuring, repeating or testing supernatural events.... No matter how many stones intelligent design throws at the theory of evolution, the only alternative it presents for the development and diversity of life ... is a miracle, an abrupt appearance, an act of supernatural creation. That, by itself, establishes intelligent design as a religious argument, not a scientific argument, for the creation of biological life that cannot be taught to public school students. (*Kitzmiller v. Dover Area School District*, 9/26/2005)

This is classic boundary-work: Selective characteristics are attributed to science for purposes of distinguishing it from a "lesser" knowledge-producing activity. Inside the rhetorically constructed boundaries of science, one finds several cultural landmarks. Science is based on data, laboratory work, observation, measurement, and consensus among all scientists over provisional explanations of natural phenomena. Outside the boundaries of science, Rothschild said, one finds divine miracles, supernatural events, and religion. The features that Rothschild attributed to science may or may not correspond to what actually goes on in laboratories or peer-reviewed journals "first time through," and that is not really the point. The boundaries of science he constructed in court are later representations that cannot be analyzed in terms of their accuracy but rather only in terms of their immediate practical and strategic utility for plaintiffs' interests in getting discussion of intelligent design out of Dover High School science classes.

For sociologists, there is no absolute cultural space for "science," nor are the boundaries around that space universal or transcendent (or, in some sense, epistemologically necessary). Clashes involving scientific knowledge are unending. In the Dover trial, boundaries became discursive weapons used by adversaries to pursue their goals at that episodic moment, in that specific place, amid that particular clash (with its long path-dependent history). Of course, those people defending the legitimacy of intelligent design as part of the school science curriculum did their own boundary-work. Patrick Gillen, attorney for the defense, observed in his opening remarks:

> Intelligent design theory is really science in its purest form, the refusal to foreclose possible explanations based on the claims of the dominant theory or the conventions of the day. ... It shares the attitude of those who worked in the field of quantum mechanics, who posited the wave-particle duality, despite the fact that to some it smacked of supernaturalism. ... Dover's modest curriculum change embodies the essence of liberal education, an education that frees the mind from the confines, the constraints, the conventions of the day and, in so doing, promotes the curiosity, the critical thinking, the quest for knowledge that has served our country so well. (*Kitzmiller v. Dover Area School District*, 9/26/2005)

Gillen's challenge is to draw the cultural boundaries of science so that intelligent design appears to have a defensible location inside. Notice that the defining features of science are vastly different from those deployed by Rothschild for the plaintiffs.

To Gillen and the defense, science is about openness to new and even untested theories, resistance to dogma (this was seen as ironic by those for whom intelligent design is religion!); science is about curiosity and critical thinking. Is science therefore to be defined by the knowledge that scientists accept as legitimate because of its observable and measurable support, or is science to be defined by its process of endless searching and skepticism of received wisdom? Emphatically, it is not the job of sociologists to answer this question (as though they, like Popper, could become referees for the endless contest of deciding who and what is really scientific). In identifying the cultural boundaries such as those that were drawn and redrawn at the Dover intelligent design trial, sociologists are to watch how boundary-work serves the professional interests of scientists seeking to retain exclusive and authoritative jurisdiction over the domain of natural truths and how it serves the interests of Dover parents and school board members seeking to insert their beliefs about biological origins and diversity into the science curriculum.

The opening remarks by Rothschild and Gillen at the *Kitzmiller v. Dover* intelligent design trial launched one recent skirmish in the centuries-old clash of knowledge involving science and religion. Adversaries constructed different boundaries and spaces for science as they sought a legal mandate for including intelligent design in school science classes—or for excluding it. To be sure, in choosing to watch science as it takes place in the Harrisburg Federal Building and Courtroom, the sociologist arrives at a conclusion that is hardly a startling revelation. The very idea of a law court compels the architecturally orchestrated co-presence of adversaries in the spatial presence of a judge (or jury) who will produce a binding verdict. It is easy to miss the critically important role of this *place*, a courtroom, in fomenting a clash between science and religion. But there are plainly other places where science happens (religion, too), and they are typically located, designed, and built in a way that minimizes the likelihood of a clash of knowledge.

Boundary-work does not happen all of the time, nor in all places. Depending on where one happens to look, the boundaries between realms of knowledge—or, more broadly, between cultural systems—exist politely, never triggering the clash and contestation so heated inside the Harrisburg courthouse during *Kitzmiller v. Dover Area School District*. Only occasionally (and in identifiable conditions like courtrooms) do the cultural boundaries between science and non-science become the object of actors' explicit discursive practices, destabilized (or defended) in the pursuit of credibility and legitimacy. Only occasionally does science become a contingently constructed space, with boundaries that are only as durable as their immediate discursive utility in contests for power and control. For the rest of the time, nobody bothers to ask, or *needs* to ask whether this is science or not. So, what preempts boundary-work? What averts the clash of knowledge? In places other than those built purposefully to force adversaries to confront their differences face-to-face, the boundaries of science (or religion) are settled. They are so thoroughly institutionalized and stabilized that "everybody knows" what science really is. The line between science and other domains of culture is treated unproblematically, as though it were a given, as though it were fixed for all working purposes.

Simply put, what social conditions obviate the need for people even to wonder about the cultural boundaries of science, much less dispute them? What allows scientists (and others) to get on with their lives with the presumption that everybody already knows what science is and is not. To find answers, sociologists must look in other kinds of places where science occurs, in buildings that ensure the institutionalization and routinization of cultural boundaries that just "are" (rather than being contingently constructed rhetorical objects of contestation). Science assumes a more settled state (for example) at Stanford's Clark Center, a spectacularly beautiful research facility that, in the materiality of its bricks, glass, and mortar, answers the question "What is science?" even as the people who work there (and those looking in) have little warrant to ask.

The Clark Center was hailed as "Laboratory Building of the Year" in 2004. Its 245,000 gross square feet cost about $147 million and has a maximum occupancy of 700 workers. The building consists of three separate wings, rectangular on the outside perimeter but concave on the inside to create an open-air courtyard. It has three stories and a basement. All of the spaces facing the courtyard have floor-to-ceiling windows and are rimmed with balconies so that anybody can see what is going on in every lab or office. Two wings are mainly for wet-bench experiments; the third is for computational work. The cavernous research spaces have an industrial feel because they are almost completely open and because all of the utilities (electricity, for example) drop down from a fully exposed four-foot ceiling zone. Unseparated by walls or even partitions, members of one research group spread into the next. Inside the vast open laboratory spaces, all of the benchwork, cabinetry, desks, large pieces of heavy equipment (such as a centrifuge) are on wheels so that they can be moved around easily in response to rapidly changing research projects and patterns of collaboration between scientists. Even office pods are on wheels so that they can be situated (temporarily, of course) near or far from benches where experiments are furiously underway. Some of the lab benches are conspicuously painted bright yellow (black is the norm) to signify "hotel space" for visiting scientists, who often come from other universities or corporations for short periods. The Clark Center is located strategically at the intersection of Stanford's other buildings for basic life science research, engineering, and medicine.

According to Stanford's public-relations machine, the Clark Center is "the vanguard of a new era," a "radical lab planning arrangement … that is designed to remodel the landscape of scientific and technological research" (Adams, n.d.). Stanford President John L. Hennessy called it "a building whose architecture mirrors our vision of the groundbreaking work that will go on there" (Baker, 2003). Chemist Tom Wandless said that "it's an experiment in social engineering" (Hall, 2003, p. 6).

What kind of science is the Clark Center trying to engineer by virtue of its strategic location, stunning design, and cutting-edge infrastructure? An answer to that question exposes a different cultural boundary of science—not religion, but politics. The intersection of science and politics has the potential to be as contentious as the boundary between science and religion. However, in contrast to the fracas in the Harrisburg courthouse, all seems calm and agreeable inside Stanford's Clark

Center. Nobody there seems troubled by the difference between "is" (science) and "ought" (politics). Everybody seems too busy with their experiments to worry much about the stuff of politics: allocating scarce resources, planning for a good society, and satisfying the diverse interests of stakeholders through compromise or sheer power. Researchers rarely discuss the larger political implications of their work. Avoiding a clash, they suspend consideration of exactly where the line is to be drawn between science and politics. The search for new knowledge about nature occupies the full attention of those working at the Clark Center, who seem to have little time for politics.

Actually, the Clark Center is full of politics, but in this place politics coexist peacefully with science, and boundary-work recedes almost invisibly into the implicit. Politics are inscribed in the walls and floors of the Clark Center, where they are very difficult to discern through the lenses of architectural beauty or technical efficiency. (There is no question that the place is gorgeous and that it works.) The Clark Center is indeed engineered, just like any other technological artifact, so its visions of a good society, its power and interests, its desires and fears, get built into the architecture of the place (Winner, 1986; see Gieryn, 2002). There is no clash of knowledge in this laboratory building, even though both science and politics are present inside. Scientific research is front stage, and politics lurk in the wings, so deeply embedded in backstage materiality that nobody seems to notice the potential for contention or the need for boundary-work.

Whose politics drove the design of the Clark Center? Which political ambitions were translated into the architecture and materiality of this building, which, in its spatiality, provides one built-in map of the borderlands between science and politics? What is the political definition of science such that this laboratory, the Clark Center at Stanford, becomes the perfect place to pursue it? John H. Marburger, III, is Science Advisor to President George W. Bush, Director of the Office of Science and Technology Policy, and, incidentally, holder of a Ph.D. in applied physics from Stanford University (1967). Marburger's many speeches and interviews offer a cartographic display of the intersection of science and politics. Specifically, he creates a space for science targeted at specific identifiable political goals. Speaking before the Council on Governmental Relations, Marburger (2006) addressed the future of the American research university. He acknowledged that these institutions are in a "volatile state" and face an "indefinite future." These unpredictable circumstances, especially in the absence of "central planning," increase the need for "flexibility to respond to changing conditions." Referring specifically to university investments in new buildings, Marburger stated bluntly that the U.S.'s "decentralized system" for funding research creates "competitiveness for research grants in a target area" and that appealing new facilities can lure "outstanding new faculty who can attract new grants." He proposed a "collective business model" for research universities, warning that there "are bound to be losers" in the anticipated "tilt toward private sector research" that will bring about a "much stronger link between economic productivity and research." At a time when there will be an "increasing intensity of competition for a large and expanding but finite federal research funding," Marburger looked to increasing the share paid for by the "private sector, particular by industries that benefit from technologies that build on the scientific products of the universities." In a speech at Rensselaer Polytechnic Institute, Marburger (2003) emphasized the "entrepreneurial" nature of scientific research these days, encouraging scientists to "take risks" and noting that the "commercialization" of fields like nanotechnology offer "natural bridges to interdisciplinary collaboration."

It is difficult to miss the free-market logic that drives Marburger's politics of science. Research is very nearly reduced to the quest for technological innovations that will restore America's global market competitiveness. That faith in market competitiveness colors Marburger's thinking about the future of the American research university, which must struggle for scarce funding by adopting flexible, interdisciplinary, problem-oriented (or targeted) research agendas in an entrepreneurial spirit and by producing knowledge commodities with commercial potential. To be sure, this emphasis on the commodification of science is only one among many ways to trace the boundary between science and politics. For example, Marburger says little about the need for central planning to insure that science is directed toward the public good and that taxpayers' support of research should produce new ideas and products that are subsequently made available freely (or cheaply). Differences of opinion on whether science is a public good or a profitable commodity could, under certain conditions, elicit the same kind of intense debate and boundary-work that took place over intelligent design in the Harrisburg courthouse.

But that debate does not happen inside Stanford's Clark Center, where the settled boundaries between science and politics are so deeply embedded in "necessary" architectural and infrastructural designs that nobody notices them anymore. The Clark Center was conceived of and built to maximize the values and goals expressed in Marburger's rhetoric. Marquee architect Norman Foster was hired to design a signature building to lure scientists with proven abilities to obtain grants. The open floor plan is the pinnacle of decentralization and flexibility, for space can be opened up or shut down quickly and cheaply in response to whatever line of inquiry suddenly seems promising commercially. There are no walls to divide scientists into discipline-bound silos. Yellow "hotel" lab benches welcome transients from industry, benefiting both Stanford and corporations through the immediate exchange of ideas and interests. The Clark Center stands at the junction of pure and applied research, proximate to work in basic sciences, engineering, and medicine.

Nobody asks about the alternative visions of science that got left outside the Clark Center. The building itself provides one ready and convincing answer to the question of what science is, an answer well aligned with the current political economic structure of resource flows on which Stanford, and certainly every major research university, depends. When Norman Foster and Bio-X scientists initially sat down at the design table to sketch out this new jewel of a lab, there was surely abundant boundary-work, for the group faced decisions about what science is (and what its intersections with political economy are). The architect's atelier, like the courtroom, is a place that invites contestation over cultural boundaries that remain in an unsettled state until ground is actually broken for a new building. But now that the Clark Center has been constructed and occupied, it provides only answers (no longer explicit boundary-work). They are visible in the kind of research projects undertaken there, in the patterns of collaboration and communication within the Center's spaces, in the grants coming in, and in the patents going out. No one has time to ask about the cultural boundaries of science and politics. They have been built-in, settled ... with no clash of knowledge.

The places that people build shape the social practices inside. To explain why the juxtaposition of knowledge does not always result in a disputatious clash, one must ask *where* cultural systems encounter each other. Some buildings, through their physical design, ornamentation, and the symbolic understandings associated with them, engender passionate conflict over cultural boundaries. Courtrooms and perhaps architects' studios are examples. Other places bury the potential for argument in arrangements of brick and mortar that settle the boundaries and remove them from explicit discursive struggle. Place segregates contention from calm, allowing the settled boundaries of science to coexist with never-ending clashes over where lines between realms of knowledge should be drawn.

References

Adams, A. (n.d.). The James H. Clark Center is open for interdisciplinary business. Stanford University School of Medicine. Retrieved January 15, 2007, from http://med.stanford.edu/research/centers/archive/clark.html

Baker, M. (2003, October 29). Clark Center, 'nucleus for a range of new research' opens. *Stanford Report*. Retrieved January 15, 2007, from http://news-service.stanford.edu/news/2003/october29/xopening-1029.html

Gieryn, T. F. (1983). Boundary-work and the demarcation of science from non-science: Strains and interests in professional ideologies of scientists. *American Sociological Review, 48*, 781–795.

Gieryn, T. F. (1999). *Cultural boundaries of science: Credibility on the line*. Chicago, IL: University of Chicago Press.

Gieryn, T. F. (2002). What buildings do. *Theory and Society, 31*, 35–74.

Hall, C. T. (2003, October 20). Formula for scientific innovation: Omit walls. Design of Stanford's Clark Center fosters interdisciplinary research. *San Francisco Chronicle*, p. A6.

Kitzmiller v. Dover Area School Board (2005). In the United States district court for the middle district of Pennsylvania, Case No. 04cv2688, Judge Jones. Retrieved January 15, 2007, from http://www.talkorigins.org/faqs/dover/kitzmiller_v_dover.html

Marburger, J. (2003). National priorities in science and technology policy. Presidential Lecture Series, Rensselaer Polytechnic Institute, Troy, NY, November 14. Retrieved January 15, 2007, from http://www.ostp.gov/html/11-14-03%20jhmRPILecture.pdf

Marburger, J. (2006). Emerging issues in science and technology policy. Address delivered before the Council of Governmental Relations, Washington, October 26. Retrieved January 15, 2007, from http://www.ostp.gov/html/jhmCOGR10-26-06.pdf

Swidler, A. (1986). Culture in action: Symbols and strategies. *American Sociological Review, 51*, 273–286.

Winner, L. (1986). Do artifacts have politics? In L. Winner (Ed.), *The whale and the reactor: A search for limits in an age of high technology* (pp. 19–39). Chicago, IL: University of Chicago Press.

Chapter 4
Actors' and Analysts' Categories in the Social Analysis of Science

Harry Collins

Actors' and Analysts' Categories

Let it be accepted that sociological explanation must begin with the perspective of the actor. The causes that give rise to anything that can be seen as consistent actions among actors turn on regularities as perceived by the actors first and the analyst second. If the analyst brings the idea of a mortgage to the study of the life of a tribe living in the Amazon jungle, then nothing consistent will emerge, for the tribe does not organize its existence around the idea of mortgage. Likewise, if the analyst brings the idea of the poison oracle as used by the Azande tribe to the study of life in Western Europe, nothing consistent will emerge, for western Europeans do not organize their lives around the divination of witches by administering poison to chickens. Insofar as analysts are going to develop categories of their own—analysts' categories—to do the work of explanation, those categories will have to be built upon actors' categories.

But where do actors' categories end and the analysts' categories start? In other words, given the idea of the double hermeneutic, there is still a choice to be made about the role of the two components. I want to start by thinking about how we make the choice in science studies, particularly in the analysis of scientific controversies.

Actors and Analysts in the Study of Science

From the very beginning, science studies have been beset with the problem of how much science you need to know to be able to analyze science. "Science warriors," such as Alan Sokal, insist that to understand the causes that lead scientists to switch from one belief to another one must have a complete grasp of the science itself. As Giles (2006) reports in reference to this author:

> Sokal says he is struck by Collins's skills in physics, but notes that such understanding would not be enough for more ambitious sociology research that attempts to probe how cultural and scientific factors shape science. "If that's your goal you need a knowledge of the field that is virtually, if not fully, at the level of researchers in the field," says Sokal. "Unless you understand the science you can't get into the theories." (p. 8)

P. Meusburger et al. (eds.), *Clashes of Knowledge*.

Some historians of science work this way, and in the early days there was tension, largely dissipated now, between this kind of historian and those sociologists who were less technically proficient (on the broad relations between analysts and science itself and how these lead to different outcomes, see Collins, 2004a, pp. 783–799. For myself, after discovering that my kind of work could in fact be done without a technical understanding of the science sufficient to be able to contribute to the field—and it may well not have turned out that way—the conceptual tension has been finally resolved with the idea of "interactional expertise." Interactional expertise is a deep understanding of the language of the science being studied, and it is gained through immersion in the discursive world of the actors without immersion in their physical world (see Collins, 2004a, pp. 745–782; Collins, 2004b, 2008; Collins & Evans, 2007; Collins et al., 2006; www.cf.ac.uk/socsi/expertise).[1] Interactional expertise is the ability to talk the science even if one cannot do the science.[2]

But if the idea of interactional expertise resolves the problem of how much scientific grasp one needs to be able to do the kind of work my colleagues and I do, it does not provide a rule for when part one of the double hermeneutic gives way to part two. I think that many of us have simply glossed over this problem for years. We have not even noticed that it exists. Certainly, I can say as a participant in the field of science studies that I had never really noticed that it existed until this very chapter began to take shape. In more concrete terms the problem goes as follows: Suppose I am analyzing the way Joe Weber's claims about the discovery of gravitational waves came to be rejected (see, for example, Collins, 1975, 2004a). I immerse myself in the discourse of gravitational wave physics and learn to understand all the arguments that were used by the actors in their debates with one another. Most of these arguments will be reproduced in my account of the ending of the controversy. But at a certain point I will say to the actors: "You don't really understand how your world works. I understand it better." This point becomes clear when the actors tell me things such as are contained in the following remark made by Richard Garwin, Weber's most influential critic in the 1970s:

> I do not consider you "a trained observer of human behavior," so far as concerns the gravity wave field. Science and technology move ahead through advances in instrumentation and publication of results. Not through gossip or "science wars" or deep introspection about what the other guy is thinking or what one is thinking oneself. (Personal communication, March 13, 2001)

This is one of the most important actors in the world that I take it upon myself to describe, and he is telling me that I do not understand that world—his world. My

[1] My apologies for the overwhelming number of self-citations in this paper, but it is a matter of working out the consequences of a brand new program.

[2] I am grateful to Peter Meusburger for reminding me that this point reflects a similar debate in the case of the arts. In Collins and Evans (2007) we do discuss the relationship between the sciences and the arts. We claim that an important difference is that the consumer's role in the legitimation of knowledge is bigger in the arts than in the sciences, so the nonperforming critic also has a more legitimate role from the outset. In science the right of the outsider to comment critically on the content of a science is much harder to establish.

response, of course, is that it is he who does not understand his own world. Here, then, I have thoroughly abandoned the actors' perspective. So far as I can see, I have never before even noticed that what I was doing was abandoning the actors' perspective and substituting my own contradictory perspective. I have certainly never thought about how such a move could be justified, and I do not know of any existing discussion of the matter.

Nevertheless, I think it is clear that social analysts of science do the right thing when, at a certain point, they abandon the account of the world provided by the actors and substitute their own account. Without this move there would be very little substance to the sociology of scientific knowledge (SSK). What can one say in favor of the move in the absence of a fully worked-out justification? Firstly, as in any science, justification must come to an end and one simply has to do the analysis and look to the outcome as its own justification. This is not an excuse to stop thinking about the problem, but it is a reason not to give up one's apparently successful scientific practice as soon as one has found a philosophical or methodological difficulty. (Collins and Yearley [1992] suggest that paralysis, reminiscent of the fate of logical positivism, follows from too much self-reflection on method.)

Secondly, the move is consistent, not arbitrary: The move is always made at roughly the same point in the investigation with roughly the same consequences, so it does not have a *post hoc* self-serving look about it. Furthermore, the move grows out of epistemological considerations. It is meant to show how the world of science works; the move is not designed to reach any particular substantive conclusion in the case of any particular scientific controversy.[3] The consistency of the move, irrespective of the contents of the science, holds out the hope that some good systematic way of accounting for the move in epistemological terms might one day be found.

Thirdly, as time has gone by, many of the actors themselves have begun to recognize the value of this kind of sociological perspective on their world. They do not have to become sociologists or buy into the entire sociological perspective to see that valuable understandings do emerge from this sociological approach. One might describe the situation in terms of interactional expertise and contributory expertise.[4] Social analysts superimpose their contributory expertise in the analysis of scientific controversies on their interactional expertise in the world of the actors. Sometimes this involves contradicting the actor's understandings of their own world. Those of the actors who have acquired a degree of interactional expertise in the social analyst's world have begun to see the point. They find that, at the very least, social analysts' contributory expertise can enrich their understanding of their world, if not overturn it. The positive reaction of many of the actors, painfully won over the years, is reassuring.

---

[3] There are some observers who think the goal should be to strengthen the voice of the weaker party in a scientific dispute. But because it is not always clear who the weakest is, and because sometimes the weak will become strong as time passes, the prescription cannot be applied consistently even if it could be justified, and I have never seen a justification (see Ashmore, 1996; Collins, 1996; Scott et al., 1990).

[4] Contributory expertise is the expertise needed to make a practical contribution to the subject under study. Interactional expertise is the expertise required to talk fluently about it.

Using Symmetry Asymmetrically

So far, it has been "discovered" that a necessary move from accepting actors' categories to rejecting actors' categories is always made in the standard analysis of scientific controversies under SSK and that this move has, as far as can be seen, never been analyzed, warranted, or even remarked upon in SSK (my apologies to those who have made remarks that I have overlooked).[5] Now I raise my gaze from the way individual scientific controversies have been analyzed under SSK to broader patterns of analysis in our analytic community. What has the SSK analysis of scientific controversy been used for?

It seems to me that the SSK analysis of scientific controversy has been most widely used to "deconstruct" scientific authority. Trevor Pinch and I used it this very way in the widely read first volume of *The Golem* series (Collins & Pinch, 1993/1998). There we wrote about "levelling the scientific terrain" and analytically conquering the forbidding peaks of scientific authority such as "Mount Newton" and "Mount Einstein" (p. 141). All this was to be accomplished by showing that the logic of science was not so far removed from the logic of everyday life. In other words, we were weakening scientific authority by imposing the analysts' world on that of the actors.[6] Our typical move was to take a scientific episode that appeared to have been closed by the overwhelming weight of theory and experiment, open it up again, and show that, insofar as it was ever closed, it was closed by "nonscientific" means. The license imparted by this kind of analysis for contemporary policy issues is to show that controversies declared closed by "the scientific authorities" are still open. The viewpoint of those with dissenting voices is reexamined and shown not to have been defeated according to the standards of science. A protoexample from chapter 2 of the first of *The Golem* series is the falsification of the widely accepted notion, enshrined with authority in most physics textbooks, that the Michelson-Morley experiment of 1887 showed the speed of light to be a constant. This is incorrect. In fact it took about 40 years for it to become widely established that the speed of light was a constant. As late as the 1930s papers were being published and prizes awarded for work showing that it was not a constant. If Trevor and I had been around in, say, 1920 and had encountered scientists arguing that Einstein must be right because the speed of light had been shown experimentally to be a constant, we would have been able to reply: "No, it has not—there is still a controversy about that." If some scientist had said to us: "That's not a real controversy, just a few mavericks who refuse to accept Einstein in the face of all the evidence," we would

---

[5] The move is essentially the same thing as the kind of imperialism that many anthropologists try to avoid. As I understand it, evaluation of actors' worlds is considered incompatible with analysis of the actors' worlds (though the anthropologist can, of course, express an opinion in his or her time off as it were). Peter Meusburger points out that a similar debate has gone much further in the study of religion.

[6] Cleverly using our rhetorical nous to describe our project as merely "display[ing] science with as little reflection on scientific method as we can muster" (Collins & Pinch, 1993/1998, p. 2)

have said: "You don't understand your own world." There are also more recent cases in which this kind of logic has been put to use:

1a Scientists working for the plant-breeding industry say that genetically modified crops are safe to plant, but the analyst says that, no, there is still a scientific controversy going on about that.
1b According to the British government, scientists say that Bovine Spongiform Encephalopathy (BSE) cannot be transmitted to humans, but the analyst says that, no, there is still a scientific controversy going on about that.
1c The British government says that scientists have shown the combined mumps, measles, and rubella (MMR) vaccine to be safe, but the analyst says that, no, there is still a controversy going on about whether the MMR vaccine causes autism in some children.

So far so good, but a warning alarm is sounded by the existence of another set of arguments:

2a The U.S. government says that scientists cannot agree about whether global warming is a real threat. The analyst says that, yes, they can and that those people who say it is not a real threat are a small minority who should be ignored and that they are serving the interests of the government.
2b The tobacco industry says that scientists cannot agree about whether tobacco causes lung cancer. The analyst says that, yes, they can and that people who disagree are a small minority who should be ignored and that they are serving the interests of the industry.
2c The motor industry says that scientists are unable to agree over whether lead in the atmosphere caused by exhaust emission from cars lowers the IQ of children. The analyst says that, yes, they can and that people who say it does not are a small minority serving the interests of the motor industry.

The two types of argument are set out in Table 4.1.

Unlike the move toward disagreeing with the actors' categories at a certain point, which is consistent with saying that each controversy studied was settled by *nonscientific* means, this argument sometimes goes one way and sometimes another. Only sometimes does the analyst overrule the actor's categories and say the controversy was not closed "scientifically." At other times the analyst says that, scientifically speaking, the controversy is closed. As with the other type of case, there is no explicit justification for the way the relationship between actor and analyst goes, but this time it is more worrying. If an argument sometimes goes one way and sometimes another, without an external justification, it can be self-serving. It could

Table 4.1 Two types of argument used by social analysts when looking at controversies

TYPE	Government/industry claim	Social analysts' claim	Social analysts' conclusion
1	Consensus over P	Significant disagreement	No consensus over P
2	No consensus over P	Disagreement insignificant	Consensus over P

be that analysts decide in advance whose side they are on and then choose the direction of the argument according to the way they want it to come out. My impression as a participant in the broad field of science and technology studies (STS)[7] over recent years is that there is some self-serving in the way the argumentative strategy is chosen. If my impression is correct, STS is changing from a discipline concerned with the nature of knowledge to a social movement concerned with defense of the powerless and support for green issues, with the epistemology being plugged in each time in whichever way gets the political job done best.

My impression as a participant could be backed up by a survey of the content of recent presentations at conferences and of recently published papers. I suspect that such a survey would reveal that the large majority of such papers and presentations argued in favor of environmental issues and the like, the relationship between analyst and scientific actors sometimes going one way, sometimes another, depending on the analyst's preferred political stance. It is a case where Max Weber's entreaty to confirm adequacy at the level of meaning with causal adequacy in explanations would be useful. Unfortunately, I do not have the data to hand or the means to collect it, but we can do a little more analytical work before we finish.

The analysis seems to show another consequence of a shift from a concern with scientific knowledge to a concern with policy.[8] The additional consequence is that policy concerns and social-epistemological concerns have a different logic when it comes to the analysis of scientific controversies. To do scientific knowledge work, one always reopens scientific debate; to do policy work, sometimes one reopens what people take to be closed, and sometimes one closes what people take to be open. That is a consequence that we should embrace. But how might we embrace it while avoiding the charge of being *post hoc* and self-serving?

It is often useful to start with an extreme case and work back to less clear-cut and more difficult examples. Let us begin, then, with "green-ink letters." Scientists (and here I can include myself), often receive letters from those who believe they have found a fundamental flaw in the theory of relativity or have developed some new all-inclusive theory of the universe. After I publish something in the science news journals, or after one of my books is reviewed in the scientific press, I often receive three or four such items, recognizable by certain characteristics. They are often rich in mathematical symbolism and, in the old days, when they came by post, they were mostly characterized by peculiar formatting. They might be written in green ink, or closely typed on both sides of the paper with no margins, or written on lined paper with no introduction or conclusion. These communications are what I call green-ink letters. Among them there may be one or two that really are of world-shattering importance, but for practical purposes one has to assume that they are not. Again, in practice there is insufficient time (even if one had the competence)

---

[7] STS is a much broader study of science, technology, and its relation to society in which sociology of scientific knowledge is subsumed.

[8] Collins and Evans (2002) try to put this shift on a systematic footing.

to track down the flaws in each case to the point where one could be certain that there was nothing in them.

I believe that someone who felt it interesting could take any one of these communications and apply the tools of SSK to reveal that the kind of process scientists use to reject green-ink letters is not scientifically pure or decisive. It would then be possible to resurrect the logic of any one of the claims, showing it to be not completely unworthy of consideration. This effort would be a perfectly proper and revealing exercise in SSK (though perhaps only suitable as a training exercise nowadays since we know in advance that it could be done and that therefore the outcome would not count as a discovery but merely a display of competence). The point is, however legitimate and valuable an exercise it would be in SSK, it would not be a proper and valuable exercise in science policy. Today's routine technical decisions cannot be made on the basis that relativity might be wrong and that all the money going into orthodox research based on relativity should be put on hold until the matter is resolved. This case is one where the policy analyst has to say that, even though some people want to say that the argument about relativity is still open, it is "really" closed. It is a Type 2 case, not a Type 1 case.[9]

Or consider the following imaginary example. I wake up one morning and decide that cancer is caused by drinking coffee. I point out the long-term correlation between the massive increase in coffee-drinking in my country and the increase in cancer as the recorded cause of death. Furthermore, there is a rough correlation at the level of whole societies between high consumption of coffee and expenditure on cancer therapies. I send out a press release, and the newspapers pick it up and run the story. Members of the public report a number of incidents in which someone was diagnosed with cancer a few months, or years, after they increased their consumption of coffee. After a short while, the existence of a connection between coffee-drinking and cancer becomes widely accepted. The relationship between coffee-drinking and cancer becomes part of the actors' perspective. Many coffee growers are bankrupted, and their laborers, deprived of wages, become weak and ill.

Does such a train of events constitute a scientific controversy? Once more, the sociologist of scientific knowledge could treat the matter symmetrically and use it to explore the ways in which one scientific idea gets promulgated and another does not. Such an investigation would show that there is no certain proof that coffee

[9] This, incidentally, is one of the problems for the position adopted by Brian Josephson as expressed at the Heidelberg conference that is the source of this volume. Josephson has discovered that the arguments deployed by his scientific colleagues to dismiss the likes of cold-fusion or homeopathy are not up to the standards of the canonical version of science. He correctly infers that there remains a small chance that there is something in them. What does not follow, however, is that the chance is large enough to make them worth pursuing. Josephson is right to fault the rhetoric in the dismissal of these maverick claims but wrong in drawing the conclusion that the associated controversies are not over for nearly all practical purposes. If it is true that absence of evidence is not evidence of absence, it is equally true that absence of disproof is not disproof of absence.

ingestion does not contribute to the onset of cancer. But, again, for policy purposes, this case cannot be treated as Type 1 but must be treated as Type 2. For policy purposes, there is no scientific controversy here. For policy purposes we have to say that this kind of thing is not a scientific controversy or anyone would be able to start a scientific controversy whenever they wished.[10]

How might one argue that these two cases are Type 2 rather than Type 1 given that it is known from the analysis of scientific knowledge that every controversy can be reopened? It is a hard problem. Perhaps one solution, admittedly not a very satisfactory one, is to look at origins. When it comes to policy, the charge "genetic fallacy" should no longer be treated as a decisively damning criticism. For policy purposes the origin of a controversy can play a part in the decision-making process. In the case of green-ink letters, it is precisely the origin that warns against taking their policy implications too seriously. In the case of coffee and cancer, it is again origins. "I wake up one morning and decide ..." is the giveaway.

The invocation of origins can be used only in extreme cases, however.[11] The courts typically assess the credibility of expert witnesses by references to their origins, and, of course, as SSK has shown, scientists do this on a regular basis as a means of finding a resolution to the problem of "experimenter's regress" (see Collins, 1992, for example). It is not analysis of origins of this relatively subtle kind that I am putting forward as a possible policy choice. That subtle kind of discussion of credibility belongs *within* a scientific controversy. The decisions being looked at in this context are about whether a scientific controversy even exists. It is being suggested that a certain scientific credibility is required in order to provide a license for starting a scientific controversy. A certain amount of scientific work by a reasonably credible scientist has to be done before the analyst should say, "This is a scientific controversy." Consequently, the analyst can sometimes say, "This is not a scientific controversy" and press the case that not just anyone should be able to dream one up.

Of course, even a dreamed-up medical controversy, if it gets going, has to be dealt with. As every social scientist knows, to deal with such a thing one must start by understanding the actors' perspective. In this case it might well be discovered that the actors do believe there is a genuine scientific problem and will treat denials by the authorities as a cover-up intended to save, say, the coffee industry (in this case). People whose first reaction is to take the side of the powerless will side with the actors and plug in the social epistemology in the style of a Type 1 controversy. Such a response implies that there is a scientific justification for the abandonment

[10] Of course, they would need power with the media, but in this context I am dealing with the logic of the analysis of science, that is to say, the logic of how sociologists exercise power as analysts. Whether it is significant power is another matter. We must always write our books and papers on the assumption that they will have the same political impact as, say, Marx's *Capital*. It is worth noting that so little is known about the detailed causal structure of the medical world that there is ample scope for dreaming up medical controversies, and it seems to happen quite frequently.

[11] Thanks to Martin Weinel for pointing out the possible confusion discussed in this paragraph.

of coffee-drinking. It is one thing to understand the actors in order to subvert their actions and persuade them that they are partaking in a moral panic rather than a matter of serious concern; it is another thing to justify their actions on the grounds that the scientific controversy is "real."

Going back to the controversies summarized in Table 4.1, I find that it looks very much as though case 1c—the debate about the MMR vaccine—is rather like the imagined coffee-cancer controversy. The difference is that the person who "woke up one morning and decided that autism was caused by the MMR vaccine" was a medical doctor who had published results showing that the measles virus might be associated with autism. The doctor first announced the connection between autism and the combined MMR vaccine *per se* at a press conference. However, even he recommended that parents continue with the single-shot measles vaccine. There seems to be no scientific evidence, only anecdotal reports by parents, that the MMR vaccine *per se* was associated with autism. These observations are sociological, not scientific. One need know nothing of the biology of the gut, the nature of vaccines, the etiology of autism, or the methods of epidemiology to recognize that this case was not a "real" scientific controversy. An analysis of the origins of the controversy is good enough. Case 1c, then, should really be case 2d. In the absence of a full survey, this case does seem to illustrate the dangers inherent in the situation represented by Table 1. It does appear that the position adopted by some social analysts was self-serving, and it does suggest that social studies of science might be becoming a social movement rather than a discipline concerned with epistemology.

It is fitting that a contribution to a book emerging out of a workshop held in Heidelberg, the home of Max Weber, should be concerned with the tension surrounding the idea of understanding the actors' perspective. For decades I have described myself as an interpretative sociologist, never quite noticing the violence I was doing to actors' categories as an integral part of my analysis of how science "really works." But I think I now see that Weber was right and that interpretation alone is not enough. I have discovered the aforementioned violence in my own work. I have at one point suggested a survey as a useful supplement to the *verstehende* (interpretive) method. In this case it would be useful if adequacy at the level of meaning were topped up with a bit of causal adequacy. But most important, I have argued that in the case of policy analysis of the sciences, as opposed to knowledge analysis, a still more brutal choice has to be made between groups of actors. This choice cannot be avoided if sociology is to be practiced as the kind of science for which Weber argued and if social analysts of science are to avoid slipping into the politically appealing rhetoric that he warned against. The appeal simply to take the actors' perspective merely sidesteps this necessary choice.[12] The next task is to find a better way to separate scientific controversies into their two types—a way that does not refer to the political desirability of the outcome. I have suggested that an examination of the origin of a controversy is one such means, but this is just a start.

[12] The abdication of responsibility is still more clear in cases like that of AIDS treatment in South Africa (see Weinel, 2008).

References

Ashmore, M. (1996). Ending up on the wrong side. *Social Studies of Science, 26*, 305–322.

Collins, H. M. (1975). The seven sexes: A study in the sociology of a phenomenon, or the replication of experiments in physics. *Sociology, 9*, 205–224.

Collins, H. M. (1992). *Changing order: Replication and induction in scientific practice* (2nd ed.). Chicago, IL: University of Chicago Press.

Collins, H. M. (1996). In praise of futile gestures: How scientific is the sociology of scientific knowledge? *Social Studies of Science, 26*, 229–244.

Collins, H. M. (2004a). *Gravity's shadow: The search for gravitational waves*. Chicago, IL: University of Chicago Press.

Collins, H. M. (2004b). Interactional expertise as a third kind of knowledge. *Phenomenology and the Cognitive Sciences, 3*, 125–143.

Collins, H. M. (Ed.) (2008 in press). Case studies in expertise and experience [Special Issue]. *Studies in History and Philosophy of Science, 39*(1).

Collins, H. M., & Evans, R. (2002). The third wave of science studies: Studies of expertise and experience. *Social Studies of Science, 32*, 235–296.

Collins, H. M., & Evans, R. (2007). *Rethinking expertise*. Chicago, IL: University of Chicago Press.

Collins, H. M., & Pinch, T. J. (1998). *The Golem: What everyone should know about science* (New ed.). New York: Cambridge University Press. [Original work published 1993.]

Collins, H. M., & Yearley, S. (1992). Epistemological chicken. In A. Pickering (Ed.), *Science as practice and culture* (pp. 301–326). Chicago, IL: University of Chicago Press.

Collins, H. M., Evans, R., Ribeiro, R., & Hall, M. (2006). Experiments with interactional expertise. *Studies in History and Philosophy of Science, 37*, 656–674.

Giles, J. (2006). Sociologist fools physics judges. *Nature, 442*, 8.

Scott, P., Richards, E., & Martin, B. (1990). Captives of controversy: The myth of the neutral social researcher in contemporary scientific controversies. *Science Technology and Human Values, 15*, 474–494.

Weinel, M. (in press). Primary source knowledge and technical decision-making: Mbeki and the AZT debate. *Studies in History and Philosophy of Science*.

Chapter 5
Science and the Limits of Knowledge

Mikael Stenmark

> Science, it is often said, is the religion of our era. Where once we expected priests to give us insight into the nature of the cosmos and of human existence, now we look rather to men, and sometimes women, in white lab coats. Where once public expenditure in the service of deeper truth might have taken the form of mighty cathedrals, today it will be found in cyclotrons and gene-sequencers. (Dupré, 2001, p. 4)

The overwhelming intellectual and practical successes of science have led some people to think that there are no real limits to the competence of science, no limits to what can be achieved in the name of science. There is nothing outside the domain of science, nor is there any area of human life to which science cannot successfully be applied. A scientific account of anything and everything constitutes the full story of the universe and its inhabitants. Or, if there are limits to the scientific enterprise, the idea is that science, at least, sets the boundaries for what we humans can ever know about reality. This is the view of *scientism*.

The historical roots of scientism can probably be traced to the Enlightenment with its ideology of progress and perfectibility. Perhaps its best-known historical advocate is the French social philosopher Auguste Comte (1798–1857) and his attempt to create a religion based on science—the "Religion of Humanity" (Comte, 1830/1988). Another interesting and far-reaching attempt to have science take over many of the functions of religion and thus itself become a religion was undertaken by the German chemist and Nobel Prize-winner Wilhelm Ostwald (1853–1932). He argued for science as an *Ersatzreligion*—a substitute religion (Ostwald, 1912; see Hakfoort, 1992).

Yet many different forms of scientism have emerged over the last three centuries. In recent decades a number of distinguished natural scientists, including Peter Atkins, Richard Dawkins, Carl Sagan, and Edward O. Wilson (as well as philosophers like Daniel D. Dennett and Patricia Churchland), have advocated scientism in one form or another. Besides receiving a number of prestigious scientific prizes and awards, these scientists have sold an enormous number of books. The views of these scholars have been discussed in newspapers and broadcast on radio and television. If scientism has been around for a while, the great impact these advocates of scientism have had on popular Western culture is new. They have brought not only science but also scientism right into the living room of ordinary people.

P. Meusburger et al. (eds.), *Clashes of Knowledge*.

What I do in this essay is give an overview of different kinds of scientism and argue that scientism (at least in some forms) is a kind of religion, a religion that is worth taking seriously. In particular, I focus on scientism and the question about the limits of scientific knowledge. I argue that scientism is a problematic position to take, one that in the end ought to be rejected.

Scientism

Let me start by giving some recent examples of spokespersons for scientism. The American philosopher and Darwinist Daniel C. Dennett (1995) writes that Darwin's dangerous idea (that is, evolution by natural selection) bears "an unmistakable likeness to universal acid: it eats through just about every traditional concept, and leaves in its wake a revolutionized world-view, with most of the old landmarks still recognizable, but transformed in fundamental ways" (p. 63). "Darwin's dangerous idea is reductionism incarnate, promising to unite and explain just about everything in one magnificent vision" (p. 82). The biologist Richard Alexander (1987) talks about the most recent discoveries in evolutionary biology as the "greatest intellectual revolution of the [twentieth] century" (p. 3). He claims (just like Dennett) that these insights will have a profound impact on the self-image we humans have—to such an extent that "we will have to start all over again to describe and understand ourselves, in terms alien to our intuitions" (p. 3). Richard Dawkins (1989) is equally, if not even more, optimistic when it comes to what modern biology can deliver. He claims that we have "no longer … to resort to superstition when faced with the deep problems: Is there a meaning to life? What are we for? What is man?" (p. 1). According to him, science, particularly biology, is capable of dealing successfully with all these questions.

In an essay entitled "The limitless power of science," Peter Atkins, Professor of Chemistry at the University of Oxford, advocates the "omnicompetence of science" and believes that "science, with its currently successful pursuit of universal competence … should be acknowledged king" (Atkins, 1995, p. 132). Lastly, the philosopher Patricia Churchland (1986) writes, "In the idealized long run, the completed science is a true description of reality, there is no other Truth and no other Reality" (p. 249).

For a philosopher of religion these ideas about science as a universal acid that eats through just about everything, as a complete explanation, as an answer to the existential questions of human beings, or even as the king of all are very fascinating. These ideas are fascinating because in them the traditional borderline between science and religion is erased. The scientific project becomes a religious or a world-view project communicating the idea that science is what can help humanity solve its sorrows and problems. Let us humans put our trust in science because it can save us from evil.

But what is scientism more exactly? It is not all that easy to define it, but it could be said that someone is an advocate of scientism if he or she believes that everything (or at least as much as possible) could and should be understood in terms of science. (Be aware that I am using the notion of science in the restricted way that is common in English but not in German or Swedish. Thus, it covers only the natural sciences and those areas of the social sciences that are highly similar in methodology to the natural sciences.) It is assumed that there is something problematic, inferior, or even irrational about the activities or enterprises that could not be understood in such a way. In a demon-haunted world, science is the candle in the dark. To spread the light of science to the "pagans" or to the "unenlightened" is therefore a part of the mission of the scientistic faith.

Another concept that could be used in this context is "scientific expansionism." It explains quite well what it is all about, namely, that the advocates of scientism believe that the boundaries of science (i.e., the natural sciences) could and should be expanded in such a way that something not previously understood as science can now become a part of science. Science can answer a lot more questions than what people have thought was possible. Scientism, in its most ambitious form, can be defined as the view that science has no real boundaries—that it will eventually answer all empirical, theoretical, practical, moral, and existential questions and will in due time solve all genuine problems humankind encounters.

How exactly the boundaries of science should be expanded and what more precisely it is that is to be included within science are issues on which there is disagreement. Some promoters of scientism are more ambitious in their extension of the boundaries of science than others. That is to say, they are all scientific expansionists, but in different ways and to different extents.

Perhaps the best-known form of scientism expresses a particular idea about the boundaries of knowledge. Epistemic scientism says that only science can confer genuine (in contrast to apparent) knowledge about reality. The only kind of knowledge people can have is scientific knowledge. Everything outside science is taken as a matter of mere belief and subjective opinions. Consequently, the agenda is to strive to incorporate as many other areas of human life within the sciences as possible so that rational consideration and acquisition of knowledge can become possible in these fields as well. Of course, it is not difficult to understand that a person holding this epistemological view believes that everything (or at least as much as possible) could and should be understood in terms of science. After all, what we humans cannot understand and explain in terms of science is something we cannot know anything about.

I call advocates of scientism "science believers." The reason why is not that I want to draw a contrast between what we humans believe and what we know and thereby indicate that science believers only believe these things but do not really know them. The reason why I use the term science believers is that I want to highlight the "believe in" rather than the "belief that" aspect of belief. The point is that these people put their *faith* in science; they put their *trust* in science; they *rely* on science. Science is, in Paul Tillich's terminology, their "ultimate concern" (Tillich, 1951, pp. 11–12).

When science fulfills a particular task in the lives of people, it becomes their religion in a functional sense. When science is what guides them in their lives, when it is what helps them to deal with their existential and moral questions, then science becomes their religion. They become science believers. So the first answer to the question of why one should understand scientism (at least in some forms) as a religion is that science in these people's lives fulfills a similar function as traditional religion does in the lives of religious believers. The second answer is that some science believers explicitly claim that science could and should replace traditional religions.

Edward O. Wilson, a professor in biology, now retired from Harvard University, is perhaps the best recent example of someone claiming that science could (and actually must replace) traditional religion. He does not call this view scientism but scientific naturalism, scientific humanism, or scientific materialism. Wilson (1978) believes that scientific naturalism "presents the human mind with an alternative mythology that until now has always, point for point in zones of conflict, defeated traditional religion" (p. 192). He also adds that the best scientific theory on which to base one's scientific mythology or religion is evolutionary theory: "The evolutionary epic is probably the best myth we will ever have" (p. 201).

Nevertheless, let me ask whether scientific naturalism satisfies the requirements for being a religion or a worldview, whether it can fulfill that particular function in human life. I have argued elsewhere that for something to be a worldview or a religion in the functional sense it must satisfy certain requirements (Stenmark, 1995, pp. 235–268). Is scientism in the form of scientific naturalism able to do this?

A worldview must fulfill at least two tasks. First, it must structure and make reality intelligible (the theoretical function of a worldview). That is, it must to some degree make the world a cosmos and determine the place of human beings in it, and it must state what is of value in life. Second, a worldview must concretely guide people in how they should live their lives, how they should deal practically with their existential experiences of, for instance, meaninglessness, suffering, guilt, and love and their interpersonal relationships (the regulative function of a worldview). These two requisites hold because believing in a worldview is not just a matter of seeing the world in a particular way but also of choosing a way of living. Scientific naturalism is able to fulfill the theoretical task. It can give its adherents a map of reality. It can say where we human beings fit in and what the central values of our existence are. It is less certain whether scientific naturalism can concretely regulate people's lives in the way traditional religions have been able to do. Wilson (1978) seems aware of this problem. He writes that the "fatal deterioration of the myths of traditional religion [has led to] a loss of moral consensus, a greater sense of helplessness about the human condition and a shrinking of concern back toward the self and the immediate future" (p. 195). Scientific naturalism must face this challenge. It must supply people with a new myth powerful enough to overcome these destructive consequences of the deterioration of traditional religious myths. It must be able to provide a faith by which people actually could live, not only with a theoretical map of reality. Wilson thus suggests

> a modification of [traditional] scientific humanism through the recognition that the mental processes of religious belief—consecration of personal and group identity, attention to charismatic leaders, mythopoeism, and others—represent programmed predispositions

> whose self-sufficient components were incorporated into the neural apparatus of the brain by thousands of generations of genetic evolution. As such they are powerful, ineradicable, and at the center of human social existence. ... I suggest further that scientific materialism must accommodate them on two levels: as a scientific puzzle of great complexity and interest, and as a source of energies that can be shifted in new directions when scientific materialism itself is accepted as the more powerful mythology. (pp. 206–207)

However, it is not possible now to predict the form that religious life and rituals will take as "scientific materialism appropriates the mythopoeic energies to its own ends" (Wilson, 1978, p. 206). Wilson admits that here lies at least the present "spiritual weakness" of scientific naturalism. It lacks the "primal source of power" that religion for genetic reasons is hooked up with. It is bereft of this power partly because the "evolutionary epic denies immortality to the individual and divine privilege to the society" (pp. 192–193). Moreover, scientific naturalism will "never enjoy the hot pleasures of spiritual conversion and self-surrender; scientists cannot in all honesty serve as priests" (p. 193). But Wilson, nevertheless, believes that a way exists to divert the power of religion into the service of scientific naturalism, even if the future will have to show how exactly this will be done. So it is clear that science is a kind of religion for some people and that they also believe science should be the religion everyone should adopt; there is a missionary incentive. Therefore, it seems as if scientific naturalism or scientism can be, or at least have, the potential to become a full-fledged religion, even if it does not have some of the attributes of traditional religions.

The Scope of Scientific Knowledge

Let me now focus more closely on epistemic scientism and the issue about the scope of scientific and human knowledge. Atkins (1995), in his argument for the limitless power of science, claims that

> there is no reason to suppose that science cannot deal with every aspect of existence. Only the religious—among whom I include not merely the prejudiced but also the underinformed—hope that there is a dark corner of the physical universe, or of the universe of experience, that science can never hope to illuminate. (p. 125)

Thomas Nagel, an agonistic, disagrees. He rather thinks this "overuse" of science to explain everything and anything (which Atkins exemplifies) has to do with a "fear of religion," that is, one's wanting atheism or naturalism to be true and hoping that there is no God but being made uneasy by the fact that many intelligent and well-informed people are religious believers (Nagel, 1997, pp. 130, 133). This fear of religion, or what he calls the "cosmic authority problem" (p. 131) is not a rare condition, and Nagel's guess is that it is responsible for much of the scientism and reductionism of our time. Be that as it may (Nagel might after all be religious in the sense of being underinformed), is there any reason to doubt the omnicompetence of science—the idea that science tells or eventually will tell everything there is to know about reality?

There are two significantly different strategies that are currently used within the Academy to argue against scientism, to argue that there is reason to doubt the omnicompetence of science.

1. The Postmodernist-Relativist Strategy (or the Social Constructivist Strategy)

The first strategy is to argue that all scientific theories are merely social constructions. Which theory prevails at any time is just a matter of who has the power and what is in fashion; there is no objective truth to be found in science. What is true about scientific theories is also, of course, true about scientism—the grandest metanarrative of them all! Paul Saulson is a scientist who identifies this kind of criticism against scientific knowledge claims (and by extension also against epistemic scientism). He writes, "I believe in the Church of Science. ... We scientists fit the profile of true believers who are convinced we have privileged access to the Truth, and who are confused about and suspicious of people who want to treat our belief system as [merely] a social phenomenon" (Saulson, 2001, p. 227).[1]

Let me give just one example of this kind of criticism. Julie Hopkins (1996) expresses a radical feminist and theological version of the postmodernist-relativist strategy when she states that

> traditionally in the patriarchal West, truth has been considered an objective reality, to be deduced through reason and then tested empirically. Elite councils of men who wielded power in the church, or science or politics have claimed that their objectivity is God-given, corresponding with metaphysical laws. [Sharon] Welsh argues that this understanding of knowledge, far from being value-free, is a strategy undertaken in order to dominate; for so-called objective reality is in fact laden with the presuppositions and prejudices of those who hold hegemonies of power and who project these onto a fictitious *tabula rasa*. Welsh agrees with the French philosopher Michel Foucault that westerners should give up the pretension of speaking in universal and dogmatic categories and recognise that a just and peaceful future lies with a new epistemology, which she names the 'political economy of truth' in which every group, class, race, sex and religion has the right to name for themselves what is true and liberating. (p. 68)

This reasoning is a form of criticism directed at the idea of an objective science, a criticism that, if true, undermines scientism as well.

2. The Rationalist, Multiknowledge Strategy

It is the second strategy that I attempt to develop in this essay. The basic idea is to grant that science affords knowledge of reality but to argue that (a) it is merely *one* kind of knowledge that is possible to obtain and (b) these other forms of genuine

---

[1] Although Saulson writes in this way, he is probably not to be considered an advocate of scientism.

knowledge cannot be reduced to scientific knowledge. It is an argument for the plurality of human modes of knowing.

What is seriously problematic about the epistemological form of scientism is that it is in fact self-refuting (that is, it undermines itself). Epistemic scientism states that all genuine knowledge is scientific knowledge. It follows from this claim that we humans cannot *know* that scientific knowledge is the only mode of knowledge unless we are scientifically able to determine that it is. This inference is valid because science is the only source of knowledge that we have. But what methods in, for instance, biology or physics are suitable for such a task of showing that the only kind of knowledge that we have is scientific knowledge? Well, they are hardly those methods that make it possible for scientists to discover and explain electrons, protons, genes, survival mechanisms, and natural selection. The reason why is not that the content of this belief is too small, too distant, or too far in the past for science to determine its truth-value (or probability). Rather, it is that beliefs of this sort are not subject to scientific inquiry. The belief that only science can give us knowledge about reality is a view in the theory of knowledge and is, therefore, a piece of philosophy and not a piece of science. But then epistemic scientism is self-refuting because *if* the belief that only science can give us knowledge about reality is a philosophical standpoint, *then* it follows that we can never know that it is true, for the belief itself says that the only kind of knowledge we can have is scientific knowledge!

The only way around this problem is to try to reformulate the scientistic stance, but is reformulation possible? The answer appears to be "yes," and I think Atkins (1995) can help here:

> The attitude that I advocate is that the omnicompetence of science, and in particular the simplicity its reductionist insight reveals, should be accepted as a working hypothesis until, if ever, it is proved inadequate. (Atkins, p. 132)

But why should people accept this perspective as their working hypothesis? The answer that Atkins hints at is that science has been tremendously successful; it has provided insights about nature that people could only have dreamed about a couple of centuries ago. It is the success story of science that justifies the scientistic attitude.

Although one cannot prove that all genuine knowledge is scientific knowledge, the assertion that it is could and should be taken as a working hypothesis because of the success story of science. Other kinds of knowledge might therefore exist. Unlike the "hard-core scientism" discussed above, this "soft-core scientism" does not deny that other kinds of knowledge might exist. Instead, it tends to place the burden of proof on those who claim that there are kinds of knowledge other than scientific knowledge. The advocates of soft-core scientism merely maintain that people should be suspicious of all human knowledge claims that are not scientific and apparently not reducible to scientific knowledge claims. This form of scientism appears not to be self-refuting. It does not require people to know that all genuine knowledge is scientific knowledge and thus does not pose the impossible task of explaining how this knowledge could be obtained by scientific means. The advocates of soft-core scientism would rather maintain that it is a rational belief or a justified working hypothesis; it is rational or justified because of the success story of science.

Nonscientific Modes of Knowing

For the sake of argument let me assume that the burden of proof falls on those people who believe that there are forms of knowledge other than the scientific kind. Let me address the issue of whether there are things that it is reasonable to assume we humans know but that are not scientifically knowable. What kind of other valid epistemological activities (if any) are there besides science?

I claim that the record of scientific success is enough to justify belief that science, by and large, gets at the truth about the world and therefore generates knowledge, but it justifies no more than that statement because there are other reliable forms of knowledge. Scientific knowledge even presupposes the existence of some of these other kinds of knowledge. If we did not know certain things already, we would not be able to obtain any scientific knowledge at all. What do we need to know to be able to do science? (See Stenmark, 2001, pp. 18–33)

Suppose I am (like Darwin was) thinking about why a new species comes into existence and an old one exits the scene, and suppose I come up with the idea of natural selection. In this particular case, I observe a herd of zebras in which there seems to be variation of running speed. I formulate the thought that perhaps the running speed is inherited and that it makes a difference to survival or reproduction—and I am fully aware that I have these thoughts. This kind of knowledge about my thoughts (and emotions) could be called "introspective knowledge." Not only do I have introspective knowledge (knowledge about what I am thinking of), but it seems like it is *more certain* than scientific knowledge. I am more certain that I have this thought than that running speed is inherited and that it makes a difference to survival or reproduction of the zebras. This certainty is highlighted by the fact that it makes no sense to question my belief that I am now thinking about the running speed of zebras by asking, "Mikael, are you certain about this? Is it not rather the case that you are thinking about food?" This question, however, is reasonable when it comes to my ideas that running speed is inherited and that it makes a difference to survival or reproduction of the zebras. What grounds do I have for claiming these things?

Moreover, I must know that I have these kinds of thoughts to be able to do science at all. If I cannot know that *I* have this thought (that I am thinking about the running speed of zebras), how am I going to be able to develop complex theories about the relationship between variation, fitness, and heritability? That achievement does not seem to be possible. But not only do I have to know that I have these thoughts to do science, I must also over a period of time *remember* that I have had these thoughts as well as all the observations I have made about the running speed of individual zebras. To be able to develop and test a scientific hypothesis against a certain range of data, scientists have to be able to remember, for instance, the content of the hypothesis, the previous test results, and, more fundamentally, the fact that they are scientists, and the location of their laboratories. Their scientific knowledge presupposes knowledge based on memory.

But could not introspective knowledge and knowledge based on memory be reduced to scientific knowledge? No, because if scientific knowledge presupposes introspective knowledge and knowledge based on memory, then one *first* must *know* these things to be able to do science. Reduction of scientific knowledge to introspective knowledge and knowledge based on memory is therefore not possible. Moreover, it does not seem reasonable to reduce knowledge claims that are more certain to knowledge claims that are less certain.

I also argue that knowledge about social reality is something that science cannot confer, and it is a kind of knowledge few people would on reflection deny that human beings have. Let me give you an example of what I mean by knowledge of the social world. I am not talking about the social sciences but merely about common sense knowledge, or everyday life knowledge. Suppose I go into a café in Stockholm and sit on a chair at a table. The waiter comes and I utter a fragment of a Swedish sentence. I say, "Kan jag få en öl, tack?" The waiter brings the beer and I drink it. I read a book and notice a Coca-Cola sign on the wall and a car outside the window. I leave some money on the table and leave. This sequence of actions sounds simple, but as John Searle (1995) has pointed out, its metaphysical complexity is truly staggering. Moreover, its significant features fall outside science.

Notice that the language of physics, chemistry, or any other of the natural sciences cannot capture the features of the description I have just given. There is no physical-chemical description adequate to define "café," "waiter," "sentence in Swedish," "money," or even "chair" and "table," even though café, waiters, money, chairs, and tables are physical phenomena. Because no physical-chemical description can be given of these social phenomena, no scientific knowledge of the social world exists. But we humans know them; a large chunk of our knowledge is of the social world we inhabit! Where science can see only masses of metal in linear trajectories, we can see a car. Where science can see only cellulose fibers with green and gray stains, we can see dollar bills. Moreover, I do not merely order a beer to drink. I am also reading a book in which it is written, "The only kind of knowledge we can have is scientific knowledge." Now, to put it bluntly, can science read books or, for simplicity, the two sentences "The only kind of knowledge we can have is scientific knowledge" and "Drink Coca-Cola" and thus obtain *linguistic knowledge*? More precisely, the question is whether, for example, the biologist qua biologist or the physicist qua physicist can read these texts? Can they as scientists discover (or come to know) the meaning of these sentences by applying solely the methods of biology or physics? Well, scientists can, of course, analyze the chemical laws that allow ink to bond with paper and the other things that make it possible to write these sentences. But can scientists with these methods come to know the information contained in these sentences? I must admit that I cannot even imagine what such an experiment would look like.

Let me go back to the café again. After I have read this sentence in the book, I shake my head and look up and notice once again the Coca Cola sign on the wall of the café. But then I suddenly realize that I do not merely know that "Drink Coca Cola" is written on this billboard sign; I also know that someone is trying to convince me to buy a particular product. Hence, I come to the conclusion that I have both linguistic knowledge and *intentional knowledge* (that is knowledge about people's intentions and purposes). But once again, this knowledge that someone is trying to convince me to buy a particular product is not knowledge that comes from the natural sciences.

Therefore, and contrary to what Atkins thinks, there is every reason to believe (a) that the world is bigger than the world of the natural sciences and (b) that obtainable knowledge about this bigger world cannot be reduced to scientific knowledge.

References

Alexander, R. D. (1987). *The biology of moral systems*. New York: Aldine De Gruyter.

Atkins, P. (1995). The limitless power of science. In J. Cornwell (Ed.), *Nature's imagination: The frontiers of scientific vision* (pp. 122–132). Oxford, England: Oxford University Press.

Churchland, P. (1986). *Neurophilosophy*. Cambridge, MA: MIT.

Comte, A. (1988). *Introduction to positive philosophy* [Cours de philosophie positive] (Edited with introduction and revised translation by Frederick Ferré). Indianapolis, IN: Hackett. [Original work published 1830.]

Dawkins, R. (1989). *The selfish gene* (2nd ed.). Oxford, England: Oxford University Press.

Dennett, D. C. (1995). *Darwin's dangerous idea*. London: Penguin.

Dupré, J. (2001). *Human nature and the limits of science*. Oxford, England: Oxford University Press.

Hakfoort, C. (1992). Science deified: Wilhelm Ostwald's energeticist world-view and the history of scientism, *Annals of Science, 49*, 525–544.

Hopkins, J. (1996). Radical passion. In D. Hampson (Ed.), *Swallowing a fishbone* (pp. 66–81). London: SPCK.

Nagel, T. (1997). *The last word*. New York: Oxford University Press.

Ostwald, W. (1912). Die Wissenschaft. In W. Blossfeldt (Ed.), *Der erste internationale Monisten-Kongress in Hamburg vom 8.–11. September 1911* (pp. 94–112). Leipzig, Germany: n.p.

Saulson, P. R. (2001). Confessions of a believer. In J. A. Labinger & H. Collins (Eds.), *The one culture* (pp. 227–232). Chicago, IL: University of Chicago Press.

Searle, J. R. (1995). *The construction of social reality*. New York: The Free Press.

Stenmark, M. (1995). *Rationality in science, religion, and everyday life: A critical evaluation of four models of rationality*. Notre Dame, IN: University of Notre Dame Press.

Stenmark, M. (2001). *Scientism: Science, ethics and religion*. Aldershot, England: Ashgate.

Tillich, P. (1951). *Systematic theology: Vol. 1*. Chicago, IL: University of Chicago Press.

Wilson, E. O. (1978). *On human nature*. Cambridge, MA: Harvard University Press.

Chapter 6
Science and Religion in Popular Publishing in 19th-Century Britain

Aileen Fyfe

Going Beyond the Conflict Thesis

Although it is still frequently asserted that science and religion are naturally in conflict, historians have long since demonstrated that the myth of inevitable conflict was created in the late 19th century (Helmstadter & Lightman, 1990; Russell, 1989). Rather than being rooted in a rational philosophical debate over the relative merits of scientific or theological explanations of natural phenomena, the relations between science and religion in the 19th century were grounded in social and professional structures. The myth of an inevitable philosophical conflict was a useful campaigning tool for a group of British men of science (most famously Thomas Henry Huxley and John Tyndall) who hoped to increase the cultural authority of science and the status of its practitioners at the expense of religion and its practitioners. These men found it rhetorically useful to claim that science and religion were in conflict (Turner, 1978). But, as historians have shown, the outspoken opposition to theology and organized religion that characterized many of these scientific naturalists had not been shared by previous generations of devout scientific practitioners (Brooke, 1990; Cantor, 1985; Corsi, 1988). Moreover, even in the late 19th century, many men of science retained their personal faith, most prominently, the North British group of physicists (Smith, 1998). Thus, historians have found many problems and inadequacies in the "inevitable conflict" thesis.

My aim in this chapter is to take the revisionist historiography still further. Historians have so far tended to focus on highly educated men, such as members of the respectable middle classes and professors at the universities. These members of the intellectual elite were, of course, the sort of people who might be expected to have philosophical worries about the relations between science and theology or to undergo crises of faith. The fact that so many of these intellectuals did value scientific explanations over theological ones by the end of the 19th century tells us nothing at all about what the rest of the population thought about the matter, and that silence, I believe, is a serious omission.

To address the question of what the bulk of the population might have thought about science, religion, and their interactions, one needs to think about how knowledge traveled beyond those intellectual circles. How did people outside the London elites

P. Meusburger et al. (eds.), *Clashes of Knowledge*.

come to know anything about either science or theology, let alone their relations? This question involves both a geographical and a social answer. It means thinking about people in the provinces and about people lower down the social scale. Once one begins to think in these terms, it becomes even more important to consider the social presence (rather than philosophical position) of both science and religion in 19th-century culture and society.

There is already an extensive body of work on religion in popular culture. This research has uncovered the extensive infrastructure of churches, Sunday schools, Bible classes, and missionary societies that pervaded 19th-century Britain (Knight, 1995; Martin, 1983). Britain was a predominantly Protestant country, and, despite denominational distinctions, religious faith was a key component of middle-class British life from childhood onwards, supported by church attendance, devotional reading, and family prayers (Davidoff & Hall, 2002). And for the working classes, whose limited opportunities for education were most likely to come from schools run by churches or religious charities, religious instruction comprised an intrinsic part of that very basic education (Laquer, 1976). Thus, the infrastructure of religion ensured that every child in Britain was introduced to theological explanations of the world and that these were enforced and supported by church and community throughout adult life.

Despite the great enthusiasm for religion among the Victorian middle classes, religious observation among the population at large did decline by the end of the 19th century. This trend, however, had very little to do with the rising cultural authority of science among the intelligentsia. It had far more to do with the decline of traditional community structures during rapid urbanization, the inadequacy of church provision in the expanding cities, and the growth of alternative leisure activities on Sundays (Chadwick, 1975; Williams, 1999). A few working-class radicals and socialists opposed Christianity on philosophical grounds, but the vast majority of the British population seems to have moved away from religion out of apathy.

There is also a growing body of literature on science in popular culture. Much of that research directly engages with questions of the communication and dissemination of knowledge (Fyfe, 2004; Fyfe & Lightman, 2007; Secord, 2000; Topham, 2000). At no point in the 19th century were the sciences supported by anything equivalent to the well-established system of churches, schools, and missions that promoted theological knowledge. Science did not have churches or missionaries, nor was it even a usual school subject. People fortunate enough to get secondary schooling were likely to be drilled in the classical languages. That great majority who spent only a year or two at a charity-run school or sporadically attended Sunday school once a week in the local church were unlikely to learn much more than basic literacy skills and a smattering of Christianity.

Thus, most people who knew anything about the sciences would probably have gained it from their own informal reading, attendance at lectures, or visits to museums or exhibitions. For most of the 19th century, these options were the main ways in which most people would encounter the sciences (if they did so at all). That is why I believe it is so important to study the manner in which the sciences were presented in these popular formats. Their relative importance declined only once science became a routine part of school education at the very end of the century. Although a full story of the public dissemination of the sciences should also include public lectures, exhibitions, museums, and botanic gardens, I concentrate in this chapter on printed matter. From the mid-19th century onwards, print was easily the most effective means of disseminating knowledge more widely—both socially and geographically. As one commentator remarked in 1853, the press "has long been the rival of the pulpit, and is now, if we mistake not, in the wide range of its influence, far ahead of it" (Pearson, 1853, p. 473).

Outline of the Argument

The heart of this chapter examines the relative abilities of Christian and secular science writers and commentators to engage successfully in the competitive literary marketplace and thus to bring their vision of the sciences to a wider audience. I start by outlining the role of religious organizations in making possible the flood of cheap print that contemporaries observed in the mid-19th century. Although theological knowledge, in contrast to scientific knowledge, was effectively disseminated by a variety of methods, religious organizations were nonetheless actively and innovatively involved with print culture. Although they had helped to make cheap print possible, religious groups came to feel that it threatened their Christian vision of science, and I examine how they responded. Finally, I contrast the efforts of scientific naturalists to overthrow the authority of religious explanations at the end of the century. Before beginning, however, I need to make three general points about the scope of my argument.

Firstly, I am not contrasting "science" with "religion" but am rather contrasting two alternative visions of the sciences, one Christian and the other secular. In the early 19th century, the sciences were routinely understood to be part of a theological understanding of nature. Virtually everyone who pursued the sciences at any level regarded the study of nature as the study of God's Works, which would ultimately reveal His benevolence, power, and wisdom. Anglican educational institutions, from grammar schools to the University of Cambridge, encouraged the study of nature as a suitable pursuit for Christian gentlemen, and the Scottish minister Thomas Chalmers felt astronomy to be a suitable topic for a series of sermons. Science was not something separate from religion but was widely regarded as an intrinsic part of it. This Christian vision remained strong throughout the 19th century, but it faced an increasing range of competing alternatives. For example, in the 1830s, some popular instructive publishers began to omit religion from their publications because they recognized that it was controversial and might alienate some potential customers. In the 1840s, radical atheists tried, unsuccessfully, to find an audience for a vision of the sciences that was proudly materialistic and opposed to Church and State. And by the 1860s and 1870s, the new generation of men of science, led by Huxley and Tyndall, tried to make certain that their secular vision of scientific naturalism became widespread.

My second and third points principally have to do with the development of the book trade rather than with either science or religion. The mid-19th century was the period in which print emerged as the first mass medium. At the start of the century, books (like lectures and museums) had been too expensive to be accessible to anybody outside the affluent classes. Thus, at the start of the century, print could help knowledge travel geographically but did not do much to make it available to a wider selection of social groups. By the 1840s and 1850s, this situation had been transformed as publishers came to terms with the production and distribution capacities of the new steam-powered technologies and realized the potential profits to be made from the newly literate members of the lower-middle and working classes (Twyman, 1998; Vincent, 2000). Rather than producing a small number of expensive books for a limited circle of affluent readers, some publishers began to produce large print runs of cheap works. It became possible for printed matter to reach almost all sectors of literate society (Eliot, 1994; Feather, 1988).

My third point is that British commentators were proud of their traditions of liberty and the free press. There was no censorship in Britain, although blasphemy and obscenity laws were occasionally invoked. With most of the book trade driven by purely commercial concerns, it was publishers who made the key decisions about what got published, based upon what they thought would sell profitably. In this commercial marketplace, neither clergymen nor men of science could hope to control the press. The best they could do was work with it and try to use it effectively. Given this background, I think it is significant that people supporting a Christian vision of the sciences had vastly more experience and more resources in working with the book trade than did those who hoped to promote scientific naturalism to a wider audience.

A Flood of Cheap Print

At the start of the 19th century, literacy rates were still low and printed matter was relatively expensive. Most commercial publishers made their profits from a small number of expensive books. A typical print run for a new book might be just 750 copies, and it could cost as much as 30 shillings (far beyond the reach of an artisan earning 30*s*. to 40*s*. a week).[1] By the 1850s, print runs had risen, prices had fallen, and people had started to talk about print as a mass medium. Religious organizations had been intimately involved in this transformation and thus, despite their worries about its effect, were well placed to engage in the new literary marketplace. As I show below, the scientific naturalists found it more difficult to compete.

During the evangelical revival in the late 18th and early 19th centuries, reading and the Bible had returned to central importance in British Christianity. In the absence of a state-sponsored education system (until 1870), evangelical organizations set up schools

[1] Before decimalization in 1971, Great Britain used a monetary system in which 12 pennies (12*d*.) equaled 1 shilling (1*s*.), and 20 shillings (20*s*.) equaled one pound (£1). Therefore, £1 contained 240*d*.

and Sunday schools to teach the children of the poor to read. Once literacy rates started rising, it was again evangelicals who realized that ordinary books and magazines were far too expensive for most of that readership. They set up new organizations such as the Religious Tract Society (1799; hereafter RTS) and the British and Foreign Bible Society (1804), which were dedicated to the production and distribution of cheap print (Fyfe, 2004; Howsam, 1991). The ability of the religious societies to print so cheaply was largely due to their committed dedication to enormous print runs and their willingness to forgo profits, but it was also materially assisted by their enthusiastic adoption of new technologies (principally steam-powered printing) in the 1820s. Only a few other publishers adopted these strategies in the first half of the century, and they all shared a commitment to philanthropy. The charitable Society for the Diffusion of Useful Knowledge (1826) and the private firm of William & Robert Chambers also sought to help the working classes improve themselves via suitable reading material, but both of these publishers made a point of omitting religious discussion from their publications in an effort to avoid controversy.

With the exception of these secular and religious philanthropists, most publishers took a rather longer time to wake up to the capabilities of the new technologies and the existence of the vastly expanded reading audience (Anderson, 1991; Weedon, 2003). It was around 1850 that commentators began to notice "a flood of cheap print," triumphantly declaring that "The age in which we live, is unprecedented for the cheapness and abundant supply of its literature" (Pearson, 1853, p. 478). Another writer remarked that "it is the glory of our age to have brought science and sound literature within the reach of the humblest citizen" ([Patmore], 1847, p. 124). Readers in the working classes, who had previously had little or no access to newspapers, periodicals, or books, began to have the opportunity to read: either for entertainment or for instruction in everything from politics to philosophy and chemistry.

However, this flood of cheap print created real concerns about the nature of the reading material being thus presented to readers who were barely educated. Historians of science are most familiar with the controversy surrounding the anonymous evolutionary best-seller, *Vestiges of the Natural History of Creation* (1844), but *Vestiges* was unique more for the extent of the vitriol poured upon it than for the specific faults for which it was criticized (Secord, 2000). Many commentators pointed out in despair that the most popular reading material at mid-century was novels, which were hardly calculated to improve anyone's mind. Others worried about the distortions, errors, and platitudes contained in the "popular treatises and essays without number" ([Masson], 1855, p. 166). Was it really better to have badly written introductions to the sciences than none at all? There was a real need for competent writers who could express themselves fluently and comprehensibly as well as present their subject accurately and reliably. But by far the biggest fears were about the religious sentiments—or absence thereof—in popular literature. Many middle-class thinkers were convinced that no book could be really edifying if it did not place its subject matter in a Christian perspective, and they condemned the immorality and infidelity of much cheap literature (Fyfe, 2005b).

To many commentators, it seemed all too clear that "the demon of infidelity is stalking abroad" (Pearson, 1853, p. xiv), and the press was held largely to blame for undermining the authority of the church. One enterprising author went so far as to make calculations on the subject. His 1847 pamphlet entitled *The Power of the Press: Is It Rightly Employed?* surveyed the extent of "corrupting" publications, especially periodicals, and estimated their total annual circulations as 28.9 million issues. For Christianity, he could count only 24.4 million issues (*Power of the Press*, 1847). This was a general survey of all literature, but some commentators felt that scientific publications were a particular problem. As one minister commented, "In literature and science, we have not a little in which upper and under currents of scepticism are too perceptible" (Pearson, 1853, p. 480). In particular, he thought that a "positive hostility to a pure spiritual religion, or that contemptuous disregard of it" had become "woefully characteristic of some modern works of science" (Pearson, 1853, p. 359).

For all those who had been brought up to regard science as the study of God's creation, which would illuminate His wisdom and providence, it was shocking to see publications that presented the sciences in a totally different light and deeply worrying that such publications were being so widely read. The greatest anger was reserved for atheistic—or "infidel"—publications. Perhaps the most infamous was the *Oracle of Reason*, which proudly announced itself as "the only exclusively ATHEISTICAL print that has appeared in any age or country" (*Oracle of Reason*, 1842, p. ii). Its articles on history, philosophy, and the sciences set out to disprove the Bible and used the sciences to demonstrate that the universe, and life in it, had developed without supernatural intervention. Fortunately for those alarmed by such claims, the circulation of the *Oracle* fell steadily from the 4,000 copies sold of its first number, and it ceased publication after 2 years (Chilton, 1843).

More appalling for many Christians were those works that, rather than attacking Christianity, simply ignored it. Some perfectly respectable publishers, such as W. & R. Chambers and Charles Knight, took this secular route to science publication and defended it as a way of making their publications acceptable to a far wider audience. *Chambers's Edinburgh Journal* routinely sold around 60,000 copies in the 1840s and was still being issued at the start of the 20th century (Cooney, 1970, chapter 2). As accepted family reading, it was hugely more influential than the *Oracle*'s blatant atheism. Yet the absence of Christian sentiments meant that even critics who applauded its instructive content deplored the fact that it lacked "the evangelical element—that decidedly Christian tone" (Pearson, 1853, p. 510). Secular works could be held to imply that theology was irrelevant to the study of nature.

Disseminating a Christian Vision of Science

So, what were Christians to do if they wished to maintain a theological vision of the natural world amidst growing secular and infidel competition? Censorship of the press was impossible. Rather, the answer would have to involve fighting back by providing a Christian alternative that would be competitively priced and would give consumers a choice. In contemporary parlance, these Christian works would be an antidote against the poison of atheism and secularism. The existence of the religious publishing societies gave evangelicals greater power to intervene in the world of publishing and meant that the dissemination of printed religious knowledge was not entirely dependent upon the commercial marketplace. The RTS took the lead in this battle to maintain a Christian tone in popular publishing.

The RTS had been established in 1799 to produce tracts for use by missionaries to the working classes in British inner cities. It developed into a major publishing house, issuing not just tracts but books and periodicals for adults and children. Until the 1840s, its publications had all been avowedly religious, but from 1845, it began to issue books and periodicals on general topics, such as natural history, astronomy, biography, and history (Fyfe, 2005c). The "Monthly Series" of sixpenny books began in 1845 and closed 10 years later, after one hundred volumes had been issued. The closely related *Leisure Hour* periodical was launched in 1852, at the price of one penny per week, and ran until the early 20th century. Because these publications were not intended to be devotional treatises, they rarely contained explicit discussions about the proper relationship between the Word of God and His Works as visible in nature and society. But, in keeping with the society's overall mission, they all had a Christian tone.

For instance, the geological volume, *Caves of the Earth* (1847), opened with the assertion that the earth had been "'given to the children of men' by the Divine Author of all being" (Milner, 1847, pp. 7–9). For this writer, the study of the earth clearly had a theological as well as geological rationale. Passing references to God the Creator or to God's Providence were common in these publications. For instance, the writer of *Garden Flowers of the Year* (1847) attributed spring's "bringing forth the bright verdure and radiant flowers from their wintry darkness" to "the Almighty's word" ([Pratt], 1847, p. 63). Thus, even though most of the RTS texts were similar in content to the treatises offered by secular publishers like Chambers, their readers should have been constantly aware that the sciences could and should be integrated into a theological worldview.

The RTS's works were widely welcomed by commentators, who described them as "a step in the right direction" (Pearson, 1853, p. 509). The literary periodicals praised them for being "intrinsically good" and giving an "intelligent account" (Glimpses of the Dark Ages, 1846) and for being "interesting and trustworthy" (Life of Cyrus, 1847). Although some of the praise came from explicitly religious magazines, much of it came from the general literary press, clearly demonstrating how widespread the assumption still was that the sciences ought to be presented in a Christian tone.

The RTS was not, of course, the only publisher of popular works on the sciences, and many commercial publishers were sympathetic to Christian presentations of the sciences. A substantial number of the best-selling scientific authors at mid-century were committed Christians, including several in holy orders. The Rev. Thomas Milner combined his writing for the RTS with commissions from the commercial firms of W. S. Orr and Longman & Co (Fyfe, 2005a), and the Rev. John George

Wood wrote for the entrepreneurial publisher George Routledge (Lightman, 1999). Routledge specialized in mass-market cheap books for railway travelers and others, a format that enabled Wood's *Common Objects of the Country* (1857) to sell 68,000 copies within a decade. Other ordained science writers included Ebenezer C. Brewer (*A Guide to the Scientific Knowledge of Things Familiar*, 1847), Thomas Dick (*The Solar System*, 1846), and Charles A. Johns (*Flowers of the Field*, 1853). Philip Henry Gosse (*The Aquarium*, 1854) was a lay preacher, and Margaret Gatty (*Parables from Nature*, 1855) was the wife of a clergyman. It is clear that, at mid-century, a vast amount of popular scientific writing was being presented in a Christian tone—especially at the cheaper end of the market, where religious publishing charities had a large market share.

Some of these mid-century writers were still writing in the 1860s and 1870s (and many were still selling), but they were, of course, being joined by a new generation of younger writers. If popular writing followed the trends seen in expert scientific circles, one might expect that popular writing would have become almost entirely secularized by the 1880s. Yet, although some of the new generation were committed evolutionary naturalists, it remains striking that many popular writers in Britain in the late 19th century continued to see the hand of God behind nature (Lightman, 1999, 2000).

Thus, it is clear that theological visions of the sciences were still pervasive and attractive in the British popular press long after the mid-19th century. This means that there was a growing divergence in the second half of the century between the secular writing style that dominated the writings of expert men of science, and the range of writing styles, both secular and theological, that were available in the works of the most successful popular writers. The alleged victory of scientific naturalism over a traditional Christian vision of science would not have been immediately apparent in the realms of the popular media.

Disseminating Scientific Naturalism

The continued success of Christian writers and publishing organizations meant that when Thomas Henry Huxley and his fellow scientific naturalists wished to promote their alternative vision of the sciences in the 1860s and 1870s, they faced no easy task (Barton, 1998). In their campaigns for an authoritative new profession of science in which religious thinking would have no place, they would clearly have preferred popular writings on the sciences to be secular. But, with notable exceptions, the scientific naturalists did not themselves write these popular works, and the people who did write them were not necessarily committed to writing in a style that suited the scientific naturalists. Furthermore, the reality of the publishing industry meant that it was publishers, not scientists (nor theologians), who made decisions about what would sell. As long as publishers thought there was a market for Christian works of science, they would continue to commission and publish such works. Commercial necessity meant that the publishing industry reflected popular tastes rather than the happenings in intellectual circles.

How, then, could the scientific naturalists intervene in the publishing world? When Christian thinkers had faced the same problem in the 1840s and 1850s they had been able to mount a convincing response. Not only were many commercial publishers sympathetic to Christianity, but there were several major charitable publishing houses entirely devoted to religious publishing, and they enabled Christian writers to engage directly in the book trade. Moreover, there was a vast army of committed Christians who were willing and eager to write for the cause (and a small fee). The scientific naturalists, in contrast, had the disadvantage of being few in number and having no tame charitable publishing societies ready to do their bidding.

There was, however, no shortage of commercial publishers who were willing to publish such eminent authors as Huxley. But Huxley's attitude to popular writing is revealing. Early in his career, he had been a harsh reviewer of scientific works written by nonexperts, and he had made some deeply critical remarks about the whole validity of popular writing on the sciences. In the 1860s, he had repeatedly refused to write for popular audiences, preferring to spend his time on research and on writing about that research for his fellow men of science. By the 1870s, however, Huxley had at last become convinced that an intervention was needed to bring the naturalistic vision of the sciences to a wider public—and that this would mean getting involved himself. He became involved in two publication projects, Macmillan's series of 'Science Primers' (very cheap books, especially intended for use in schools) and the Anglo-American 'International Science Series' (intended for the educated general reader). He was on the editorial board for both projects and promised to contribute a volume to both. Strikingly, however, both of his contributions were late, with his introductory volume to the 'Science Primers' series appearing almost 10 years after the rest of the volumes (Desmond, 1997; Lightman, 2007; MacLeod, 1980). Thus, although Huxley finally realized the importance of intervening in popular publishing, he himself had too many other calls on his time to be able to make those interventions effectively. In contrast, many of the Christian writers were not trying to build a scientific career for themselves. Mothers, teachers, ministers, and professional writers generally had more time available for writing and fewer fears about the impact of such writing on their scholarly reputations.

Where the scientific naturalists were successful was in their long-term strategy of institution-building. They made no attempt to build a system equivalent to the churches and missions that promulgated and endorsed theological explanations of the world, but they did become heavily involved in education. Before the Education Act of 1870, most schools in Britain were run by church groups. The Education Act supplemented these schools with nondenominational ones, in which scriptural education was optional and the governors were laypeople. Huxley and his friends were active in lobbying government to gain science a higher status in the school curriculum. Huxley also had a direct role in the implementation of the new curricula, for it was his staff at the Normal School in South Kensington, London, who ran summer schools giving teachers the knowledge and skills to teach scientific subjects that they themselves had never studied. The Normal School staff also produced school textbooks to accompany the new subjects.

By the 1880s and 1890s, the scientific naturalists were thereby able to ensure that science was taught in schools and that it was being presented from a secular perspective. This investment of time and effort in science education made certain that the next generation of children grew up far more knowledgeable about the sciences than their parents or grandparents had been. They also grew up in a society in which religion was declining in power.

Conclusions

Until the developments in the education system filtered through at the turn of the 20th century, there is no denying that religious organizations had far more effective means of disseminating their vision of knowledge. Not only was there the infrastructure of churches and church-run schools, but the activities of hordes of Christian writers, not to mention the dedicated religious publishing societies, ensured that the Christian vision of the sciences was widely disseminated through popular books and magazines. It was difficult for scientific naturalists to match this achievement, for they lacked the necessary publishing societies and did not personally have time to write many popular works. Some popular writers did choose to espouse scientific naturalism—Grant Allen was a highly successful writer on evolutionary topics—but it was not until the 20th century that they outnumbered the Christian writers (Bowler, 2006).

Thus, it seems to me highly unlikely that the population at large in late 19th-century Britain would have regarded scientific explanations as having yet replaced theological ones. Only the better educated and the determinedly self-improving would have been aware of the sorts of arguments that were taking place between scientific naturalists and theologians in the 1870s and 1880s. To understand how those arguments traveled beyond the intellectual circles requires a look at how knowledge traveled, and in 19th century Britain, that was primarily through print. And I argue that the supporters of a Christian vision of science had far more experience and success in operating through the popular press than did the supporters of scientific naturalism.

It seems clear to me that the increasing trend towards secularization in intellectual scientific circles was not straightforwardly repeated in the general population. As much as scientific naturalists would have liked to make scientific explanations as widely known and authoritative as theological explanations already were, they simply did not have control of an infrastructure equivalent to that of the Christian religions. They did make extensive use of the press, but neither they nor the religious organizations could control the press, which continued to offer a plethora of visions of the sciences long after such options had been restricted within expert science. If science has indeed replaced religion as the dominant system of knowledge, it did not do so among the general British population in the 19th century. That transition should be sought instead in the early 20th century.

References

Anderson, P. (1991). *The printed image and the transformation of popular culture 1790–1860*. Oxford, England: Clarendon.
Barton, R. (1998). "Huxley, Lubbock, and half a dozen others": Professionals and gentlemen in the formation of the X Club, 1851–1864. *Isis, 89*, 410–444.
Bowler, P. J. (2006). Experts and publishers: Writing popular science in early twentieth-century Britain, writing popular history of science now. *British Journal for the History of Science, 39*, 159–187.
Brewer, E. C. (1847). *A guide to the scientific knowledge of things familiar*. London: Jarrold.
Brooke, J. H. (1990). "A sower went forth": Joseph Priestley and the ministry of reform. In A. T. Schwartz & J. G. McEvoy (Eds.), *Motion toward perfection: The achievement of Joseph Priestley* (pp. 21–56). Boston, MA: Skinner House.
Cantor, G. N. (1985). Reading the book of nature: The relation between Faraday's religion and his science. In D. Gooding & F. A. James (Eds.), *Faraday rediscovered: Essays on the life and work of Michael Faraday, 1791–1867* (pp. 69–81). Basingstoke, England: Macmillan.
Chadwick, O. (1975). *The secularization of the European mind in the nineteenth century*. Cambridge, England: Cambridge University Press.
Chilton, W. (1843). Preface. *Oracle of Reason, 2*, iii.
Chambers, R. (1844). *Vestiges of the natural history of creation*. London: Churchill.
Cooney, S. M. (1970). *Publishers for the people: W. & R. Chambers—The early years, 1832–50*. Unpublished doctoral dissertation, Ohio State University.
Corsi, P. (1988). *Science and religion: Baden-Powell and the Anglican debate, 1800–1860*. Cambridge, England: Cambridge University Press.
Davidoff, L., & Hall, C. (2002). *Family fortunes: Men and women of the English middle class, 1780–1850* (rev. ed.). London: Routledge.
Desmond, A. (1997). *Huxley: Evolution's high priest*. London: Michael Joseph.
Dick, T. (1846). *The solar system*. London: Religious Tract Society.
Eliot, S. (1994). *Some patterns and trends in British publishing, 1800–1919*. London: The Bibliographic Society.
Feather, J. (1988). *A history of British publishing*. London: Routledge.
Fyfe, A. (2004). *Science and salvation: Evangelicals and popular science publishing in Victorian Britain*. Chicago, IL: University of Chicago Press.
Fyfe, A. (2005a). Conscientious workmen or booksellers' hacks? The professional identities of science writers in the mid-nineteenth century. *Isis, 96*, 192–223.
Fyfe, A. (2005b). Expertise and Christianity: High standards versus the free market in popular publishing. In D. M. Knight & M. D. Eddy (Eds.), *Science and beliefs: From natural philosophy to natural science, 1700–1900* (pp. 113–126). Aldershot, England: Ashgate.
Fyfe, A. (2005c). Societies as publishers: The Religious Tract Society in the mid-nineteenth century. *Publishing History, 58*, 5–42.
Fyfe, A., & Lightman, B. (Eds.). (2007). *Science in the marketplace: Nineteenth-century sites and experiences*. Chicago, IL: University of Chicago Press.
Gatty, M. (1855). *Parables from nature*. London: Bell & Daldy.
Glimpses of the Dark Ages. (1846). *British Quarterly Review, 3*, 548.
Gosse, P. H. (1854). *The aquarium*. London: van Voorst.
Helmstadter, R., & Lightman, B. (Eds.). (1990). *Victorian faith in crisis: Essays on continuity and change in nineteenth-century religious belief*. London: Macmillan.
Howsam, L. (1991). *Cheap Bibles: Nineteenth-century publishing and the British and Foreign Bible Society*. Cambridge, England: Cambridge University Press.
Johns, C. A. (1853). *Flowers of the field*. London: Society for Promoting Christian Knowledge.
Knight, F. (1995). *The nineteenth-century church and English society*. Cambridge, England: Cambridge University Press.

Laquer, T. W. (1976). *Religion and respectability: Sunday schools and working-class culture, 1780–1850*. New Haven, CT: Yale University Press.

Life of Cyrus. (1847). *British Quarterly Review, 5*, 561.

Lightman, B. (1999). The story of nature: Victorian popularizers and scientific narrative. *Victorian Review, 25*(2), 1–29.

Lightman, B. (2000). The visual theology of Victorian popularizers of science: From reverent eye to chemical retina. *Isis, 91*, 651–680.

Lightman, B. (2007). *Victorian popularizers of science: Designing nature for new audiences*. Chicago: University of Chicago Press.

MacLeod, R. M. (1980). Evolutionism, internationalism and commercial enterprise in Victorian Britain: The International Scientific Series, 1871–1910. In A. J. Meadows (Ed.), *The development of science publishing in Europe* (pp. 63–93). Amsterdam: Elsevier.

Martin, R. H. (1983). *Evangelicals united: ecumenical stirrings in pre-Victorian Britain, 1795–1830*. Metuchen, NJ: Scarecrow.

Masson, D. (1855). Present aspects and tendencies of literature. *British Quarterly Review, 21*, 157–181.

Milner, T. (1847). *The caves of the earth: Their natural history, features, and incidents*. London: Religious Tract Society.

Oracle of Reason. (1842). *1*.

Patmore, C. (1847). Popular serial literature. *North British Review, 7*, 110–136.

Pearson, T. (1853). *Infidelity: Its aspects, causes and agencies; being the prize essay of the British Organization of the Evangelical Alliance*. London: Partridge and Oakey.

Pratt, A. (1847). *Garden flowers of the year*. London: Religious Tract Society.

Russell, C. A. (1989). The conflict metaphor and its social origins. *Science and Christian Belief, 1*, 3–26.

Secord, J. A. (2000). *Victorian sensation: The extraordinary publication, reception and secret authorship of Vestiges of the Natural History of Creation*. Chicago, IL: University of Chicago Press.

Smith, C. (1998). *The science of energy: A cultural history of energy physics in Victorian Britain*. London: Athlone.

The power of the press: Is it rightly employed? (1847). London: Partridge & Oakey.

Topham, J. R. (2000). Scientific publishing and the reading of science in early nineteenth-century Britain: An historiographical survey and guide to sources. *Studies in History and Philosophy of Science, 31A*, 559–612.

Turner, F. M. (1978). The Victorian conflict between science and religion: A professional dimension. *Isis, 69*, 356–376.

Twyman, M. (1998). *Printing 1770–1970: An illustrated history of its development and uses in England*. London: British Library.

Vincent, D. (2000). *The rise of mass literacy: Reading and writing in modern Europe*. Oxford, England: Polity.

Weedon, A. (2003). *Victorian publishing: The economics of book production for a mass market, 1836–1916*. Aldershot, England: Ashgate.

Williams, S. C. (1999). *Religious belief and popular culture in Southwark, c.1880–1939*. Oxford, England: Oxford University Press.

Wood, J. G. (1857). *Common objects of the country*. London: Routledge.

Chapter 7
Reason, Faith, and Gnosis: Potentials and Problematics of a Typological Construct

Wouter J. Hanegraaff

In the late 1980s the well-known specialist of ancient gnosticism Gilles Quispel (1916–2006) edited and published a Dutch collection of articles, *Gnosis: The Third Component of the European Cultural Tradition* (Quispel, 1988/2005), to which he later added a second collective volume, *The Hermetic Gnosis through the Centuries* (Quispel, 1992). Both volumes made an attempt at tracing the history of a certain type of religious or religiophilosophical thought and practice from antiquity to the present. In the introduction to the first volume, Quispel claimed that the common factor in all these currents was the central importance of *gnosis*: a Greek term meaning "knowledge," and more specifically, a kind of intuitive, nondiscursive, salvational knowledge of one's own true self and of God. Quispel's grand thesis was that—in addition to the established churches and theologies with their emphasis on faith, and the philosophical and scientific traditions based on rationality, or reason—there had always existed a third component of the European cultural tradition (Quispel, 2005) grounded in gnosis. This tradition of gnosis, or so he argued, had always been suppressed and marginalized by the representatives of reason and faith, including modern historians, who had sorely neglected the role played by gnosis in the history of Western culture or had presented it in a very negative light. Quispel's thesis not only had its roots in scholarly considerations but reflected his personal commitment as well: in his later years he increasingly came to present gnosis as a superior spiritual wisdom, in the Dutch media he openly identified himself as a modern gnostic, and he allowed his work to be loosely associated with "new age" agendas. Unfortunately, by doing so he made it very easy for opponents to dismiss his thesis as inspired merely by apologetic agendas (for a critical analysis, see Hanegraaff, 1998, pp. 19–21).

The Hermetic Tradition

Independently from Quispel, however, scholars of religion in recent decades have increasingly begun to pay attention to the kinds of religion he referred to as belonging to the "third component of the European cultural tradition." A pioneer in this regard was the English historian Frances A. Yates. Her extremely influential 1964

P. Meusburger et al. (eds.), *Clashes of Knowledge*.

study about Giordano Bruno called attention to what she referred to as the "Hermetic Tradition" of the Renaissance, grounded in the seminal 1471 translation of a collection of texts attributed to a legendary author, Hermes Trismegistus (Ficino, 1471). These texts, known as the *Corpus Hermeticum*, were actually written in the 2nd or 3rd century and represent a current of religious philosophy that sought salvation in the attainment of true gnosis about God, the human self, and the world.[1]

It is important to realize, however, that Renaissance thinkers, including the translator of the *Corpus*, the great Florentine neoplatonist Marsilio Ficino, believed that the hermetic writings were much older. They were seen as remnants of the original, supreme, universal theology revealed by God to the ancients and kept alive through the ages by divinely inspired sages such as Hermes Trismegistus, Zoroaster, Orpheus, Pythagoras, and Plato. This ancient theology (*prisca theologia*; see Walker, 1972; Hanegraaff, 2005b) was believed to have been prophetic of the Christian revelation, and therefore the recovery of texts like the *Corpus Hermeticum* was believed to make possible a reform of Christianity by leading it back to its pristine divine origins. The enormous antiquity and, hence, authority attributed to authors like Hermes Trismegistus during the Renaissance had far-reaching effects that went far beyond the sphere of religion, touching the domains of philosophy and science as well. The *Corpus Hermeticum* emphasized the relation between knowledge of one's own self, of God, and of the world, as in this famous passage on how nature helps us understand God:

> ... you must think of God in this way, as having everything—the cosmos, himself, the universe—like thoughts within himself. Thus, unless you make yourself equal to God, you cannot understand God: like is understood by like. Make yourself grow to immeasurable immensity, outleap all body, outstrip all time, become eternity and you will understand God. ... Go higher than every height and lower than every depth. Collect in yourself all the sensations of what has been made, of fire and water, dry and wet; be everywhere at once, on land, in the sea, in heaven; be not yet born, be in the womb, be young, old, dead, beyond death. And when you have understood all these at once—times, places, things, qualities, quantities—then you can understand God. ... And do you say, "God is unseen"? Hold your tongue! Who is more visible than God? This is why he made all things: so that through them all you might look on him. (C.H. XI, 20, 22; in Copenhaver, *Hermetica*, pp. 41–42)

Such a positive approach to the world of nature as a mirror that reflects its creator correlated very well with the fact that the very same author, Hermes Trismegistus, was believed to be the author of a great number of texts on what are commonly known as the occult sciences: astrology, alchemy, and magic. These traditional disciplines had been on the ascent since the 11th and 12th centuries, when the treasures of Arabic manuscripts in these domains fell into Christian hands and were translated

[1] The standard critical edition is Nock & Festugière (1991–1992). The most reliable modern English translation is Copenhaver (1992). For an up-to-date overview of the hermetic literature and hermetism in late antiquity, see van den Broek (2005a, b, c).

into Latin, and they now came to enjoy a new prestige during the Renaissance (see Buntz, 2005; Fanger & Klaassen, 2005; Haage, 2005; Lory, 2005; Lucentini & Perrone Compagni, 2005; and von Stuckrad, 2005; on the problematics of the concept of *occult sciences*, see Hanegraaff, 2005a). Natural magic (that is to say, magic based on the use of the hidden, occult forces of nature), astrology, and alchemy were seen by many as perfectly compatible with both the ancient spiritual philosophy and Christian doctrine. Hence, a new kind of religious philosophy, or philosophical religion, began to emerge in the second half of the 15th century, a system based upon syncretic combinations of hermetism, neoplatonism, and various magical, astrological, and alchemical traditions, all of them integrated within a Christian context. To this heady mixture was added the Jewish mystical tradition known as kabbalah, which was likewise seen as reflecting the ancient wisdom. The pioneer in this regard was a contemporary of Ficino, the intellectual prodigy Giovanni Pico della Mirandola, and the complex Jewish traditions of esoteric wisdom entered the domain of Christian speculation by way of the German Johannes Reuchlin and a range of other so-called Christian kabbalists (see Dan, 1997; Kilcher, 1998; Schmidt-Biggemann, 2003; Secret, 1964/1985).

This entire cluster of traditions and developments was put on the agenda of scholarly research by Yates (1964) under the label of "the Hermetic Tradition." In claiming that it deserved serious study and had important implications for the understanding of Renaissance culture, Yates broke with a type of traditional academic attitude that was still dominant in the 1960s. It is eloquently summarized by the influential historian of science George Sarton (1975): "The historian of science [thus Sarton] cannot devote much attention to the study of superstition and magic, that is, of unreason ... Human folly being at once unprogressive, unchangeable, and unlimited, its study is a hopeless undertaking" (p. 19). This attitude of contempt had been typical of how the entire domain of the occult had been approached by scholars ever since the 18th century, and it permeates even the work of the few scholars who did study the subject in depth, such as Thorndike (1923), whose multivolume work is still indispensable. Against the drift of mainstream academic scholarship, Yates (1964) claimed that the study of hermetic magic and related currents was precisely what held the key to a correct understanding of early modern culture in general, to the true nature of the Renaissance, and to the scientific revolution in particular. In her enthusiasm, she exaggerated, and her own interpretations are now outdated in many respects (see Copenhaver, 1990; Hanegraaff, 2001). Nevertheless, she did succeed in definitely putting hermeticism and the occult sciences on the agenda of historians of science, philosophy, culture, and religion. Probably the most famous example of this shift is the case of Newton and his interest in alchemy. It is amusing to see how Westfall (1980), in his great Newton biography, apparently felt he had to assure the reader that "I am not myself an alchemist, nor do I believe in its premises" (p. 21, note 12)—as though anyone would have suspected him of such a thing—and proceeded to explain why he devoted so much space to the subject. He had no choice, he writes, for "I have undertaken to write a biography of Newton, and my personal preferences cannot make more than a million words he wrote in the study of alchemy disappear" (p. 21, note 12).

Feelings of uneasiness and fear of being criticized for discussing such subjects at all were widespread among academics until well into the 1980s and are still far from having vanished altogether.

From Hermeticism to Western Esotericism

Yates's concept of the Hermetic Tradition was limited to the period of the Renaissance, but scholars have become increasingly attentive to the fact that the traditions in question by no means ceased to exist after the 16th and 17th centuries. With respect to the 18th century, the so-called Age of Reason, one thinks of Zimmermann (1969–1979), Faivre (1973), and Neugebauer-Wölk (1999), to mention only a few of the more important scholars who, each in their own way, have called attention to the significance of hermetic, illuminist, theosophical, and related traditions for understanding the complexity of intellectual, philosophical, scientific, and religious culture during the 18th century. But whereas the role of such currents in the 15th through the 17th centuries is by now generally recognized, the issue is still very controversial with respect to the Age of Reason. For example, despite the abundance of solid research now available, large and authoritative studies such as Israel (2001) continue to present a traditional black-and-white picture of how the Enlightenment "washed away" the forces of superstition and unreason (see also Thomas, 1971). Actually, however, a greater amount of hermetic literature was published during the 18th century than in the centuries before (Kemper, 1999, p. 149), meaning that there was a market for it. In fact the eclectic phenomenon known as the *Vernünftige Hermetik* of the period (a term introduced by Zimmermann, 1969, Vol. 1, pp. 19–43) answered a widely felt need. It seemed capable of providing a kind of middle ground that was attractive to those who were convinced by the arguments of the Enlightenment but did not want to throw out the baby of religion with the bathwater of the established churches and theologies. In other words, Enlightenment Hermeticism seemed to provide a kind of natural religion that harmonized with reason and that integrated science within its theological frameworks. The importance of this dimension of 18th-century intellectual culture is now being investigated by an increasing number of scholars, along with other related aspects such as the relation of Freemasonry to the Enlightenment.

To finish my bird's-eye view of the field, there is now a cornucopia of research that documents, interprets, and contextualizes the continuation of the same cluster of ideas and traditions through the 19th and 20th centuries up to the present. Alternative currents such as mesmerism, spiritualism, modern theosophy, occultism, traditionalism, neopaganism, and even New Age (see Hanegraaff, 2005c) are now seen by many scholars as significant dimensions of the complex clashes of knowledge and cultures of knowledge in post-Enlightenment and secular culture. While earlier generations saw these currents as little more than irrational survivals and marginal pursuits with little or no relevance to what was really important in 19th- and 20th-century culture, scholars since the 1980s and increasingly since the 1990s have recognized that these currents were, on the contrary, integral parts of post-Enlightenment processes of modernization, secularization, and the disenchantment of the world. The fact that careful study of these "occultist" and related currents leads to surprising new perspectives on 19th- and 20th-century culture in general has been demonstrated independently by a variety of authors (e.g., Godwin, 1994; Harvey, 2005; Laurant, 1992; Owen, 2004; and Treitel, 2004). This new development has much to do with the decline during the 1980s of the "secularization thesis" and its assumption that rationality and scientific progress inevitably lead to an increasing marginalization of religion. Again, old-fashioned secularization and modernization narratives are being replaced by more sophisticated and more complex narratives showing how religion in fact survives and develops by variously adapting itself, often successfully, to the new circumstances of secularity, disenchantment, and so on (with reference to the New Age movement, see Hanegraaff, 1996).

The increasing recognition of the entire field of currents and traditions that I have been sketching as worthy of academic attention and important for a more nuanced understanding of Western culture is a phenomenon that has become notable especially since the 1990s. Arguably, the new openness of academics to all these traditionally suspect currents and traditions has something to do with a broadly "postmodern" *Zeitgeist*, which is instinctively critical of the hegemonic claims of the grand narratives of Western culture and emphasizes the multilayered complexity of competing discourses of knowledge and power (as suggested in Hanegraaff, 2001, 2004). Whatever the case, among the various terminologies that have traditionally been used to refer to this field, the expression *Western esotericism* has meanwhile emerged as the generally accepted label. There now exists a professional journal published by Brill (*Aries: Journal for the Study of Western Esotericism*) and an accompanying monograph series with the same publisher (*Aries Book Series: Texts and Studies in Western Esotericism*). The International Association for the History of Religion (IAHR) has been organizing symposia on Western esotericism at its congresses since 1995. The American Academy of Religion has granted protected group status to the same field since 2005. Full-time master programs are now available at the Universities of Amsterdam and Exeter, and there is reason to expect that more will follow in the coming years. There are two international professional organizations, the U.S.-based Association for the Study of Esotericism (ASE; see http://www.aseweb.org) and the new European Society for the Study of Western Esotericism (ESSWE, see http://www.esswe.org). Finally, the full range and complexity of the entire field of research is documented and demonstrated by the two-volume *Dictionary of Gnosis and Western Esotericism* (Hanegraaff, 2005c).

Gnosis and Western Esotericism

In the title of that dictionary, which represents the current state of the art in this domain of research, the precise nature of the relation between gnosis and Western esotericism was deliberately kept somewhat ambiguous. Discussion of the reason for this decision leads back to the two aforementioned Dutch volumes edited by Quispel (1988, 1992). It may sound good, even spectacular, to state that all the various currents and ideas covered under the broad umbrella of Western esotericism are united in their appeal to gnosis—but upon closer examination it appears that historical realities are not that simple. Writings by major protagonists of Western esoteric currents refer to several types of knowledge: not only to a kind of mystical gnosis as knowledge of the self and of God, important though it may often be for those thinkers, but also to knowledge based on revelation, tradition, rationality, and science. The suggestion that all the representatives of esoteric traditions are unanimous adherents of gnosis, united in their rejection of rationality and doctrinal faith, looks suspiciously like a positive reversal of traditional stereotypes according to which these traditions are "irrational" heresies, "obscurantist" superstitions, or mystical rapture (*Schwärmerei*). The concept of a "tradition of gnosis" thereby ends up (as argued in Hanegraaff, 2001) creating or confirming the artificial idea of a "counterculture" of esotericists united in their battle against mainstream religion, philosophy, and science. Modern students of Western esotericism have come to realize that this perception exists only in the imagination, not in historical reality.

Nevertheless, it must be admitted that the idea of gnosis as the third current of Western culture alongside reason and doctrinal faith has a peculiar attraction because it makes so much intuitive sense. Is it not true, one might argue, that all these "esoteric" traditions do not really fit the mold of the mainstream traditional churches and doctrinal theology ("faith") and of philosophical rationality and science ("reason"), and therefore represent a "third option"? Certainly, they have come to be perceived and presented as such, particularly since the 18th century, but again, careful research demonstrates that the distinctions break down once the sources are closely examined and put into context. In other words, the distinction between reason, faith, and gnosis is largely misleading *if it is used as a description of historical reality*. I argue, however, that it does have a valid use if understood as an analytical tool that may help distinguish different kinds of knowledge referred to by both esoteric *and* nonesoteric authors. This distinction between a historical/descriptive and analytical understanding I consider crucial.

An Analytical Typology

Against the background just sketched out, I propose an analytical typology that differentiates between three basic kinds of knowledge referred to as reason, faith, and gnosis—but from a point of view that is very different from Quispel's (1988/2005, 1992). Again, it is essential to understand that these three kinds of knowledge should not be confused with specific historical authors and currents. Quispel presented the Christian churches and theologies as based on faith, philosophy and science as based on reason, and a third current that might be called Western esotericism as based on gnosis. In contrast, I emphasize that knowledge claims belonging to any of my three analytical categories — faith, reason, and gnosis — can be found in the Christian churches and theologies, in philosophy and science, and in Western esotericism—not to mention other domains such as art and literature. What differs is only the degree of emphasis.

First, at least two questions can be asked about any claim to knowledge: Can the knowledge be checked by others, and can it be communicated to others? There are, of course, many kinds of knowledge with respect to which the answer to both questions is affirmative. For example, if I claim to know that there is an elephant in the next room, anybody interested in checking my claim can go into that room and see for him- or herself. If I claim to have discovered the solution to a complex mathematical problem or to have found a cure for AIDS, anybody with sufficient mathematical or medical knowledge can check my claim and find out whether I have made a mistake. Furthermore, not only can all this knowledge be checked by others, it can also be communicated without problems. I can use normal language to state that there is an elephant in the next room, or the languages of mathematics or biochemistry to explain my other discoveries. Any kind of knowledge that can be checked as well as communicated I propose to refer to as "reason" (and let me add that I do not attach very much importance to that particular term: if one prefers to call it "category A," that will be fine with me).

Second, there are claims to knowledge that can be communicated but that cannot be checked by others. Tradition has it that the prophet Mohammad received the Qu'ran from the angel Gabriel. The contents of this divine revelation were written down in Arabic and can thus be communicated without problems. But clear though the message may be, others cannot check whether it is actually correct, whether it has some divine origin or is just a figment of Mohammad's imagination and that of his followers. It is a question of belief. One can go into the next room to look for the elephant, but there is nowhere one can go to see whether Allah or the angel Gabriel is there (on this principle of methodological agnosticism with respect to the meta-empirical, see Hanegraaff, 1995).

It may be instructive to examine yet another example, chosen here because it concerns a figure usually categorized as an esotericist. The 18th-century visionary Emanuel Swedenborg claimed to travel to heaven and hell and talk with its inhabitants. He claimed that the spirits of the deceased visited him daily and that he could see them and talk with them. He was able to describe them in detail and relate their conversations exactly—hence, the knowledge he claimed to possess was perfectly communicable. But, of course, whether the spirits and the angels were really there remained impossible for anyone to check. This second kind of knowledge I refer to as "faith."

The third and final category is very peculiar because it consists of claims to knowledge that (like the knowledge claims pertaining to faith) one cannot check, but the contents of which cannot even be communicated—and which are nevertheless considered of the utmost importance by those who claim to have received such knowledge. Thus one ends up with the following (see Table 7.1).

To get an idea of what is meant by the third and final type, let us get back to the *Corpus Hermeticum*. Throughout this collection of texts, gnosis is not given just

Table 7.1 Typology of three basic kinds of knowledge

	Characteristic of knowledge claimed	
Analytical category	Communicable	Verifiable or falsifiable
Reason	+	+
Faith	+	–
Gnosis	–	–

like that; it is preceded by philosophical teachings, the truth of which first has to be understood by reason and then accepted as true on the authority of the teacher. The pupil receives information that can be readily communicated and understood, but the truth of which has to be accepted on faith. Characteristic of the hermetic attitude, however, is that such knowledge, though necessary as a preparation, is not enough. This aspect is explained with particular clarity in C.H. IX.:

> If you are mindful, Asclepius, these things should seem true to you, but they will be beyond belief if you have no knowledge. To understand is to believe, and not to believe is not to understand. Reasoned discourse does not get to the truth, but mind is powerful, and, when it has been guided by reason up to a point, it has the means to get as far as the truth. After mind had considered all this carefully and had discovered that all of it is in harmony with the discoveries of reason, it came to believe, and in this beautiful belief it found rest. By an act of God, then, those who have understood find what I have been saying believable, but those who have not understood do not find it believable. (C.H. IX, 10, in Copenhaver, 1992, p. 29)

Thus reason and faith are necessary prolegomena, but the actual gnosis is a gift from God and has a content that can no longer be communicated but only beheld directly by some faculty beyond the senses and reason. A particularly clear example is found in C.H. X. The pupil says that Hermes's teaching has filled him with a good and very beautiful vision that almost blinds him. But Hermes responds that the ultimate vision is even more profound:

> ... we are still too weak now for this sight; we are not yet strong enough to open our mind's eyes and look on the incorruptible, incomprehensible beauty of that good. In the moment when you have nothing to say about it, you will see it, for the knowledge of it is divine silence and suppression of all the senses. One who has understood it can understand nothing else, nor can one who has looked on it look on anything else or hear of anything else, nor can he move his body in any way. He stays still, all bodily senses and motions forgotten. (C.H. X, 5–6, in Copenhaver, 1992, p. 31)

Please note that nothing is said about the *content* of such gnosis. The text emphasizes not only that it is utterly beyond words but also that it requires the suppression of all the bodily senses. This brings me to a point that has been quite neglected by most scholars but that seems of crucial importance to me. It is that supreme knowledge or *gnosis* of the kind described in the *Corpus Hermeticum* was not just considered abstractly as a theoretical option within the overall framework of a platonic or neoplatonic metaphysics but as something that required a specific trance-like "altered state of consciousness" (ASC) accompanied by a temporary suppression of normal sensory activity.[2] That "gnosis" must be seen in the context of concrete and specific trance-like states is obvious from many of the sources if one is just attentive to it (Hanegraaff forthcoming 2009). That this dimension has seldom been highlighted can be explained most plausibly by the simple fact that most scholars of antiquity and the Renaissance are trained in disciplines like philology and philosophy, which do not have the tools needed to analyze and interpret such states. In order to really make sense of them, we will need methodologies developed by anthropology and psychology, but most of all, we must simply read what the texts are actually saying. In doing so, we need to overcome the feelings of resistance, common among historians of religions and academics generally, against giving too much attention to subjective "experiential" phenomena and to loaded concepts like trance or ASC with all the associations they evoke. Such fears are understandable enough, but they simply cannot be allowed to dictate research agendas. The historical sources themselves require that these experiential dimensions be taken seriously, so researchers need to develop methodologies that are capable of interpreting them with scholarly rigor.

Such a research agenda is only in its infancy at present. Only quite recently have academic researchers begun to take the field of Western esotericism seriously, and within that domain we have yet to begin developing theoretical frameworks and methodologies capable of making sense of the appeal to "gnosis" as subjective, experiential, noncommunicable, and nonverifiable/nonfalsifiable knowledge. If we manage to develop such frameworks and methodologies, another challenge will be to deal with the complex relation between what I have referred to as "gnosis," "reason," and "faith." As sufficiently emphasized above, and contrary to the suggestion of scholars like Quispel, we are not dealing with a simple situation of Western esotericism based on gnosis as opposed to religion based on faith and to philosophy and science based on reason. Rather, very often all three of these types of claimed knowledge turn out to be present in one and the same author or current of thought. Thus, the dimensions of "gnosis" and "faith" are by no means absent from the history of the hard sciences either, and their role and presence should be taken seriously. Likewise, it is crucial to study the role of rational discourse in the history of Western esotericism and to investigate how it relates to faith and gnosis in this particular context.

Concluding Remarks

By pursuing such lines of research, scholars may add nuance to traditional approaches of Western culture and move toward a perspective that is more complex and likely to be more adequate in making sense of historical developments. Clearly,

[2] The concept of ASC emerged in the context of LSD research in the 1960s and was introduced into academic discussion by Tart (1969/1972). Although the association of ASCs with psychoactive substances has remained strong, Tart's volume discussed various other types of ASC as well, notably those associated with the hypnagogic state, dream consciousness, meditation, and hypnosis. For a recent discussion, see Pekala and Cardeña (2000).

the intention of this short article has not been to provide definitive answers to any of the questions that have been addressed but merely to introduce an emerging field of research and point out its relevance to the general issues involved in clashes of knowledge. Most of the work still needs to be done, but it is worth the undertaking, for the implications challenge some of the most ingrained assumptions about what the term *knowledge* may have meant—and still means—in Western history and culture.

References

Broek, R. van den (2005a). Hermes trismegistus I: Antiquity. In W. J. Hanegraaff (Ed.), in collaboration with A. Faivre, R. van den Broek, & J.-P. Brach, *Dictionary of gnosis and western esotericism* (pp. 474–478). Brill, England: Leiden.

Broek, R. van den (2005b). Hermetic literature I: Antiquity. In W. J. Hanegraaff (Ed.), in collaboration with A. Faivre, R. van den Broek, & J.-P. Brach, *Dictionary of gnosis and western esotericism* (pp. 487–499). Brill, England: Leiden.

Broek, R. van den (2005c). Hermetism. In W. J. Hanegraaff (Ed.), in collaboration with A. Faivre, R. van den Broek, & J.-P. Brach, *Dictionary of gnosis and western esotericism* (pp. 558–570). Brill, England: Leiden.

Buntz, H. (2005). Alchemy III: 12th/13th–15th century. In W. J. Hanegraaff (Ed.), in collaboration with A. Faivre, R. van den Broek, & J.-P. Brach, *Dictionary of gnosis and Western esotericism* (pp. 34–41). Brill, England: Leiden.

Copenhaver, B. P. (1990). Natural magic, hermeticism, and occultism in early modern science. In D. C. Lindberg & R. S. Westman (Eds.), *Reappraisals of the scientific revolution* (pp. 261–301). Cambridge, England: Cambridge University Press.

Copenhaver, B. P. (Ed. & Trans.). (1992). *Hermetica*. Cambridge, England: Cambridge University Press.

Dan, J. (1997). *The Christian kabbalah: Jewish mystical books and their Christian interpreters*. Cambridge, MA: Harvard College Library.

Faivre, A. (1973). *L'ésotérisme au XVIIIe siècle*, Paris: Seghers.

Fanger, C., & Klaassen, F. (2005). Magic III: Middle ages. In W. J. Hanegraaff (Ed.), in collaboration with A. Faivre, R. van den Broek, & J.-P. Brach, *Dictionary of gnosis and western esotericism* (pp. 724–731). Brill, England: Leiden.

Ficino, M. (Trans.). (1471/1989). *Liber de Potestate et Sapientia Dei, Pimander*, Florence, Italy: Studio Per Edizioni Scelte (originally Treviso, Italy: Geraert van der Leye). Facsimile reproduction.

Godwin, J. (1994). *The theosophical enlightenment*. Albany, NY: State University of New York Press.

Haage, B. D. (2005). Alchemy II: Antiquity—12th century. In W. J. Hanegraaff (Ed.), in collaboration with A. Faivre, R. van den Broek, & J.-P. Brach, *Dictionary of gnosis and western esotericism* (pp. 16–34). Brill, England: Leiden.

Hanegraaff, W. J. (1995). Empirical method in the study of esotericism. *Method & Theory in the Study of Religion, 7*(2), 99–129.

Hanegraaff, W. J. (1996). *New age religion and western culture: Esotericism in the mirror of secular thought*. Brill, England: Leiden.

Hanegraaff, W. J. (1998). On the construction of "esoteric traditions." In A. Faivre & W. J. (Eds.), *Western esotericism and the science of religion: Selected papers presented at the 17th Congress of the International Association for the History of Religions, Mexico City, 1995* (pp. 11–61). Louvain, Belgium: Peeters.

Hanegraaff, W. J. (2001). Beyond the Yates paradigm: The study of Western esotericism between counterculture and new complexity. *Aries, 1*(1), 5–37.
Hanegraaff, W. J. (2004). The study of Western esotericism: New approaches to Christian and secular culture. In P. Antes, A. W. Geertz, & R. R. Warne (Eds.), *New approaches to the study of religion: Vol. 1, Regional, critical, and historical approaches* (pp. 489–519). Berlin: Walter de Gruyter.
Hanegraaff, W. J. (2005a). Occult/occultism. In W. J. Hanegraaff (Ed.), in collaboration with A. Faivre, R. van den Broek, & J.-P. Brach, *Dictionary of gnosis and western esotericism* (pp. 884–889). Brill, England: Leiden.
Hanegraaff, W. J. (2005b). Tradition. In W. J. Hanegraaff (Ed.), in collaboration with A. Faivre, R. van den Broek, & J.-P. Brach, *Dictionary of gnosis and western esotericism* (pp. 1125–1135). Brill, England: Leiden.
Hanegraaff, W. J. (Ed.), in collaboration with A. Faivre, R. van den Broek, & J.-P. Brach. (2005c). *Dictionary of Gnosis and Western Esotericism*. Brill, England: Leiden.
Hanegraaff, W.J. (forthcoming 2009). Altered States of Knowledge: The Attainment of Gnosis in the Hermetica. *The International Journal of the Platonic Tradition 3*.
Harvey, D. A. (2005). *Beyond enlightenment: Occultism and politics in modern France*. De Kalb, IL: Northern Illinois University Press.
Israel, J. (2001). *Radical enlightenment: Philosophy and the making of modernity, 1650–1750*. Oxford, England: Oxford University Press.
Kemper, H.-G. (1999). Aufgeklärter Hermetismus: Brockes' *Irdisches Vergnügen in Gott* im Spiegel seiner Bibliothek. In M. Neugebauer-Wölk (Ed.), *Aufklärung und Esoterik* (pp. 140–169). Hamburg, Germany: Felix Meiner.
Kilcher, A. (1998). *Die Sprachtheorie der Kabbala als ästhetischen Paradigma: Die Konstruktion einer ästhetischen Kabbala seit der frühen Neuzeit*. Stuttgart, Germany: J. B. Metzler.
Laurant, J.-P. (1992). *L'ésotérisme chrétien en France au XIXe siècle*. Lausanne, Switzerland: L'Âge d'Homme.
Lory, P. (2005). Hermetic literature III: Arab. In W. J. Hanegraaff (Ed.), in collaboration with A. Faivre, R. van den Broek, & J.-P. Brach, *Dictionary of gnosis and western esotericism* (pp. 529–533). Brill, England: Leiden.
Lucentini, P., & Perrone Compagni, V. (2005). Hermetic literature II: Latin Middle Ages. In W. J. Hanegraaff (Ed.), in collaboration with A. Faivre, R. van den Broek, & J.-P. Brach, *Dictionary of gnosis and western esotericism* (pp. 499–529). Brill, England: Leiden.
Neugebauer-Wölk, M. (Ed.). (1999). *Aufklärung und Esoterik*. Hamburg, Germany: Felix Meiner.
Nock, A. D., & Festugière, A.-J. (Ed. & Trans.). (1991–1992). *Corpus Hermeticum* (2 vols). Paris: Les Belles Lettres. (Original work published 1946.)
Owen, A. (2004). *The place of enchantment: British occultism and the culture of the modern*. Chicago, IL: University of Chicago Press.
Pekala, R. J., & Cardeña, E. (2000). Methodological issues in the study of altered states of consciousness and anomalous experiences. In E. Cardeña, S. J. Lynn, & S. Krippner (Eds.), *Varieties of anomalous experience: Examining the scientific evidence* (pp. 47–82). Washington, DC: American Psychological Association.
Quispel, G. (1992). *De Hermetische Gnosis in de loop der eeuwen: Beschouwingen over de invloed van een Egyptische religie op de cultuur van het Westen*. Baarn, The Netherlands: Tirion.
Quispel, G. (Ed.). (2005). *Gnosis: De derde component van de Europese cultuurtraditie*. Deventer, The Netherlands: Ankh Hermes. (Original work published 1988.)
Sarton, G. (1975). *Introduction to the history of science* (Vol. 1). New York: Krieger.
Schmidt-Biggemann, W. (Ed.). (2003). *Christliche Kabbala*. [Pforzheimer Reuchlinschriften 10]. Ostfildern, Germany: Jan Thorbecke.
Secret, F. (1985). *Les Kabbalistes Chrétiens de la Renaissance* (Rev. and expanded edition: Neuilly-sur-Seine, France: Arma Artis. (Original work published 1964.)
Tart, C. T. (Ed.). (1972). *Altered states of consciousness*. Garden City, NY: Doubleday. (Original work published in 1969.)
Thomas, K. (1971). *Religion and the decline of magic*. Harmondsworth, England: Penguin.

Thorndike, L. (1923). *A history of magic and experimental science* (Vols. 1–8). New York: Columbia University Press.

Treitel, C. (2004). *A science for the soul: Occultism and the genesis of the German modern* Baltimore, MD: The Johns Hopkins University Press.

von Stuckrad, K. (2005). Astrology III: Middle Ages. In W. J. Hanegraaff (Ed.), in collaboration with A. Faivre, R. van den Broek, & J.-P. Brach, *Dictionary of gnosis and western esotericism* (pp. 119–128). Brill, England: Leiden.

Walker, D. P. (1972). *The ancient theology: Studies in Christian Platonism from the fifteenth to the eighteenth century*. Ithaca, NY: Cornell University Press.

Westfall, R. (1980). *Never at rest: A biography of Isaac Newton*. Cambridge, England: Cambridge University Press.

Yates, F. A. (1964). *Giordano Bruno and the hermetic tradition*. London: Routledge & Kegan Paul.

Zimmermann, R. C. (1969–1979). *Das Weltbild des jungen Goethe: Studien zur hermetischen Tradition des Deutschen 18. Jahrhunderts* (Vols. 1–2). Munich, Germany: Wilhelm Fink.

Chapter 8
The Demarcation Problem of Knowledge and Faith: Questions and Answers from Theology

Michael Welker

Clashes of knowledge and faith! Currently, most people probably associate this phrase with images of narrow-minded fundamentalist creationists, or even with fundamentalist hate-preachers, with burning cars and dead bodies, with people killed by religiously motivated terrorists. The question is whether in cases like these it is really adequate to speak of a clash between realms of knowledge and whether it is adequate to attribute "faith" instead of fanaticism as the motivating force to hate-preachers and terrorist murderers. In such extreme cases, it has to be acknowledged that ideological and even pathological worldviews are beyond any areas of knowledge and beyond attitudes that can be connected with faith. In short, my first question is how seriously the term *knowledge* is taken in the title of the conference. Does the word bear a strong cognitive connotation—as in my view it should? If so, it is necessary to concentrate on problems other than the "hot" clashes between ideological and terrorist worldviews and mentalities and religiously and politically educated and civilized ones. I do not mean that the "cooler" problems I bring to attention in this text are less complicated. They do not seem to be as explosive as the examples mentioned above, but in malicious ways they are highly erosive.

In Western cultures, most of the problems with demarcating knowledge and faith seem to have been solved by efficient and peaceful modes of segmentation. I am a Christian; you are a Buddhist. He is a Jew; they are Muslims. He is a physicist; she studies German literature; and so on. These segmentations can be further refined: We are Christians, but I am a Lutheran and he is Orthodox. Or still further segmented, I am a Swabian pietist Lutheran; he is a conservative Russian Orthodox. He is not just a physicist, but an astrophysicist; she is specialized in German medieval literature. With these segmentations people operate peacefully in the spaces of knowledge and faith. He is a physicist and also an active Roman Catholic; she is interested in religious medieval literature, but she no longer practices in the Episcopalian tradition.

These refined views on fellow human beings in their participation in the spheres of knowledge and faith permit efficient and tactful, in short, adequate communication and interaction. To be sure, simple generalistic dualities such as "faith and reason" or "faith and knowledge" are still used in popular attempts to make sense of the world. In rather primitive perceptions, faith belongs to the

P. Meusburger et al. (eds.), *Clashes of Knowledge*.

areas of church, worship, and personal piety, whereas reason and knowledge belong to the areas of the academy, of research and education. However, beyond this level of very rough common-sense observations, simple dualities such as faith and reason or faith and knowledge are not capable of solving the problems of demarcating territories and boundaries of knowledge, at least not at an academic level and in environments that are shaped by nonfundamentalist religiosity. I argue that such dualities can be highly deceptive—at least in Euro-American environments shaped by Jewish-Christian traditions as well as European modernity.

The passion for insight and education characteristic of nonmystical and nonfundamentalist Jewish and Christian theology and piety discourages all attempts to supplement and support the duality of faith and knowledge with dualities such as "subjective and objective knowing," "emotional and rational attitude," or "relation to the invisible and the visible." As the long cooperation between theologians and scientists has shown, even the latter duality—the relation to the invisible and the visible—definitely collapses in the scientific realm when it comes to quantum theory (Polkinghorne & Welker, 2000).

In religious and academic communities alike, people have to deal with a set of sophisticated combinations and mixtures between trust and cognitive learning. Both groups regard themselves as "truth-seeking communities," an expression for which I am indebted to the Cambridge physicist and theologian John Polkinghorne (Polkinghorne, 1994, p. 149; 2000, pp. 29–30; Polkinghorne & Welker, 2001, pp. 139–148). The demarcation problem is thus how to differentiate between the cognitive and moral attitudes of the two communities and between their attempts to validate claims to truth and to seek an enhancement of their specific types of knowledge.

In Euro-American modernity the problem with clashes of knowledge has to be reformulated as follows: The interest in an educated faith has created a strange fusion between faith and knowledge. In order to differentiate faith from knowledge and yet preserve this fusion, modern faith has found a form that I would like to call *subjectivist faith*. It locates faith in an abstract form of self-reference. I show that this move gives faith a powerful latent pattern—and at the same time makes it empty and speechless and generates increasing self-secularization and self-banalization.

In this contribution I first analyze the structure and procedure of truth-seeking communities. In the second part, I characterize a dominant form of faith in Western societies and cultures. This form explains the strange fact that there is a high percentage of people who formally belong to the churches but that there is the constant affirmation of a spiritual hunger among people in Western societies and that they are simultaneously experiencing a strong decline in religious literacy and liturgical practice. In the third part, I try to differentiate the aims of communities in academic and in religious environments in order to describe a contrast that cannot be grasped by the unqualified duality of faith and knowledge or by its popular supplements and derivatives.

Truth-Seeking Communities

Truth-seeking communities are not to be confused with groups that announce more or less loudly that they have found the truth and now possess it.[1] Truth-seeking communities are groups of human beings who indeed raise claims to truth but who, above all, develop and practice open and public forms and procedures in which these claims to truth are subjected to critical and self-critical examination. Both the academy and many religious communities—at least in the Jewish and Christian traditions—regard themselves as such truth-seeking communities. Truth-seeking communities advance processes in which certainty and consensus can be developed, closely examined, and heightened. In doing so, however, they are to guard against reducing truth to certainty and consensus. However, truth-seeking communities also advance processes in which complex states of affairs can be made accessible in repeatable and predictable ways. In doing so, they are to guard against reducing truth to the repeatable, predictable, and correct investigation of the subject under consideration.

In my view, the path of the search for truth is adequately characterized only by the reciprocal relation between, on the one hand, the investigation and heightening of certainty and consensus and, on the other hand, the repeatable, predictable, and correct investigation of the subject under consideration. This path can be traveled only in open and public critical and self-critical communication.

People ought not to make light of the accomplishment, the value, and the blessing of truth-seeking communities, even though it is necessary to take self-critically into account the fact that these communities are always guided by other interests as well, including the search for maximum cultural resonance and for moral and political influence. They are also guided by vanity and the desire for power and control. The sober recognition that pure and perfect truth-seeking communities are rare can help balance appreciation and self-critique. It helps one be very careful about the blind self-privileging of academic work or religious communication.[2] Beware of attaching inferior value to justice-seeking communities or to communities that are committed to physical and psychic therapy and the restoration of health. There is also the obligation to respect communities that seek political loyalty and a corresponding exercise of influence, communities that seek economic and monetary success, and communities that seek to maximize public attention and resonance. It is characteristic of pluralistic societies that truth-seeking communities do not claim their truths to be absolute but rather recognize and delineate their important and indispensable contributions to the entire society and enable their contributions to be perceived in other contexts as well.

---

[1] That these mentalities can have roots in religious and scientific traditions is demonstrated by the contributions by M. Stenmark and A. Fyfe in this volume.

[2] In this volume see also E. Barker's reflections on tensions between "new religious movements" and religious institutions with long traditions.

If it is true that European modernity has equally shaped faith and knowledge in truth-seeking communities—how can religious and academic orientation be differentiated? How can each be demarcated? The success and the problem of the most dominant modern form of faith can be grasped through analysis of its inner texture, which seems to allow faith to participate in knowledge and still draw a line between faith and knowledge. In this form of modern faith, knowledge becomes self-referential and turns into an inner certainty. I call this religious, or rather quasi-religious, form "subjectivist faith."

The Structure of Subjectivist Faith and Its Religiously Destructive Power

A general understanding of faith in current Western societies is that a believing individual is utterly certain of something "wholly Other," of a "transcendent" power or authority or vaguely conceived transcendent person who at the same time, however, is intimately close (see Welker, 2004). The "beyond," the "final point of reference of creaturely dependence," the "other side" of the "founding relation of our existence" is given in an utmost, though continuously challenged, certainty. This gained, challenged, and regained certainty is called faith. This conception of faith approximates and even coincides with emphatic self-reference. The great Swiss theologian Karl Barth (1886–1968) rightly called it "indirect Cartesianism" (Barth, 1964, pp. 223–224). This indirect Cartesianism can be grasped by the formulae, "I feel somehow dependent, thus I am" and "I feel somehow dependent, thus I believe."

Because this conception of faith approximates and even coincides with emphatic self-reference, religious communication and particularly Christian theology have tried hard to differentiate this faith from all forms of self-reference. The more the inner certainty named faith has been treasured, the more all other forms of self-reference have been stigmatized and even denounced as "sin." Against this background, attempts to distinguish between innocent, trivial, and healthy forms of self-reference on the one side, and between distortive, traumatic and even demonic forms of self-reference on the other have seemed risky. A paradoxical and neuroticizing mentality has accompanied this religious form, for it has proven extremely difficult to distinguish this empty inner certainty of a wholly Other from a very simple and basic form of "pure" human self-reference that has come to terms with its inner structure, namely, that all self-reference has to include some element of difference if it wants to reach the level of experiencing "certainty."

The upside of this form of challenged and reaffirmed certainty, which can be understood both religiously and secularly, has seemed to be that nobody can escape this type of faith—at least not in cultures and among mentalities for which the self-reference of the individual is central (i.e., those belonging to typically modern world society). Because this form can appear both as a religious form and as a form of pure dialectical self-reference, it can be interpreted in a variety of ways.

This form of certainty could be used to make complex religious, moral, and metaphysical positions accessible to common sense by reducing and trivializing them and stating that in the end none of them offer anything but this dialectic of subjective immediacy and difference. For instance, it could be used:

1. Religiously as what Schleiermacher (1768–1834) called the "feeling of the utmost dependence" (*Gefühl der schlechthinnigen Abhängigkeit*; see Schleiermacher, 1821–1822/1999)
2. Philosophically as the simultaneity of self-assurance and self-challenge in the encounter with the "You ought!" of the moral law (Kant, 1788/2002)
3. Metaphysically as the dialectical unity and tension of "essence and existence" (Tillich, 1951–1963)[3]

This experience of immediacy and negation, this experience of a religious or quasi-religious certainty called faith, seems to be extremely precious and powerful. For it seems to allow religious communication to be introduced at practically any point. Nobody can escape this experience of immediacy and negation. A person trying to focus on his or her "inner self" immediately runs into this quasi-religious certainty. What is the element of the Other, whom I encounter when I try to reach the utmost depth of my inner self? Is that God? In a form that appeals to the modern mind, there seems to exist what Calvin (1559/1997), in the opening pages, called "natural awareness" and the "presentiment of the Divine."[4] To be sure, it is a culturally tamed and domesticated natural certainty. Where Calvin saw a vague awe in the face of aesthetic powers, cosmic laws, and social orders, the modern religious specimen has only a notion of the poor dialectic of empty self-awareness.

Many forms of theology, teaching, and proclamation in the classical mainstream churches have treasured this kind of abstract and empty faith very highly. They have gone to great lengths to shield this empty certainty from the discovery of its religious arbitrariness and ambiguity. They have adopted the idealist assertion that this certainty is the "foundation" of self-consciousness and the key to all epistemological and moral value and the true foundation of personality (see Welker, 2000). They have clothed this poor form with all sorts of rhetoric of "wholeness." And they have tried to reinforce the differentiation between a self-reference given by the Divine and a self-reference of purely anthropological origin. However, on the basis of the underlying theoretical construction, it has been impossible to take these attempts at differentiation and rid them of a trait of the arbitrary. As the long debates on the reflection theory of self-consciousness show, this basic dialectical relation admits of only the arbitrary definition of the "subjective and active" and the

[3] It was above all Sören Kierkegaard (1954) who repeatedly presented this form of certainty as faith and recommended it as a genuinely Christian attitude: "exactly this is ... the formula for faith: by relating to itself and by wanting to be itself, the self founds itself transparently in the power which set it" (p. 47) or "Faith is: that the self, by being itself and wanting to be itself, transparantly founds itself in God" (p. 81).

[4] See also Welker (1999, pp. 21–32) and the important differentiation between a natural awareness and a presentiment of the Divine and a "natural theology" by Pannenberg (1988).

"passive and objective" side. In reality, however, both aspects coemerge in this self-referential certainty (see Henrich, 1967, 1982; Welker, 1975).

This critical analysis of the inner texture of a typically modern form of religiosity should not lead one to underestimate its power. For this form of faith makes possible the comfortable fusion of religious and secular mentalities, of faith and knowledge. It allows one, for instance, to proceed in no time from religious to moral communication and vice versa. Above all, it is an excellent latent focus for a consumerist culture with its effort to trigger the greed-fulfillment mechanism as effectively and perfectly as possible: "already—but not yet"; "not yet—but already"; intimacy with myself, which, however, changes into the encounter with the Other; the utmost certainty and yet also the dialectical difference. Furthermore, this type of faith generously creates a religious coding of universalist mentalities. And it recursively seems to bless religious mentalities with a universalist aura. It continuously signals the message: "In a latent way, no reasonable person can be anything but religious!" If this religious form and its catalytic potential is taken seriously, it must also be made clear that it systematically prevents and discourages a content-laden and communicative piety, that it has actually driven vast parts of the Western churches into a religious speechlessness and inability to communicate.

Thus a complex religious syndrome of suffering goes along with subjectivist faith. This syndrome of suffering demands a thorough self-examination and self-criticism of modern theology and piety. Paralyzed and traumatized, the classical main-line churches in the Western industrial and information societies are obviously suffering at the beginning of the 21st century from a complex set of factors. It is that perception, not traditionalist preferences, that necessitates the examination and correction of a powerful basic form of modern religiosity.

At least five mutually reinforcing factors make subjectivist faith a power that not only blocks faith but seems to destroy it systematically. First, subjectivist faith comes in the form of a transcendental principle. It does not come—as faith should—in a form that directly animates or enlivens the communication of faith. It is individuizing and stale, a fact hidden by its universally arbitrary availability. Second, subjectivist faith comes as a necessarily empty religious form. It does not come—as faith should—in a disclosing form that gains and promotes the knowledge of God and, in its light, stimulates content-laden knowledge of self and world. Third, subjectivist faith comes as an unconditional and utmost certainty. It is a self-sufficient religious form. Although this faith can and must be activated again and again, it does not—as faith should—offer a an ordered process for passing or advancing from mere certainty to the serious individual and communal search for truth. Fourth, subjectivist faith comes as a paradoxical, self-inhibiting, even neuroticizing form in its combination of immediacy and negation. It does not promote—as faith should—the joy, doxology, and ennoblement of those who are seized by faith and who spread it. Fifth, subjectivist faith is of an escapist character. It conditions the withdrawal from expressive, festive, communicative, progressive forms of religious life and even counteracts them—as faith need not and should not.

The reason for the successful evolution of subjectivist faith must be imputed to the fact that, to many people, it has seemed to offer a simply optimal or, at least in the history of culture, a superior religiosity. Subjectivist faith is highly sensitive to and open for the concrete individual, for that person's emotional and affective forms of experience. More precisely, in principle almost completely unburdened by substantive religious matters, it is mainly concerned with the individual in his or her relation of dependence. Subjectivist faith covers the substantive side in principle through an abstract theism and totalitarian religious thought that relates everything—in fact, in a seemingly thoughtless manner—to God and God to everything.[5]

Subjectivist faith thus generates more problems than it seems to solve. In abstract theism, the question of theodicy becomes unsolvable. If God is declared to be omnipotent, how is God's goodness and love compatible with cancer, tsunamis, and concentration camps? If one is simply thrown back to empty certainty in the midst of an experience of dependence, how can this inner void be filled in a meaningful way?

Truth- and Salvation-Seeking Communities

In order to understand the self-secularizing modern type of faith, it is crucial to see that it evolved in the attempt to fuse faith and knowledge and to overcome the demarcation problem. An educated faith, a faith that seeks understanding, was embraced and cultivated or at least constructively tolerated by the university and in public religious education. Furthermore, the demarcation arrived at by subjectivist faith must be identified as a problem, as a poor solution. Self-referential religious certainty either proves to be a mere by-product of secular processes of investigation or becomes divorced from truth-seeking communities and degenerates into a mere empty certainty that can be generated again and again. The philosopher G. W. F. Hegel would have called such certainty "bad infinity" (*schlechte Unendlichkeit*).

The poverty of subjectivist faith becomes clear when one sees that it cannot incorporate two elements of faith that the search for knowledge and truth alone

[5] A whole theological network of critical encounters and movements of the 20th century collaborated in the collapse of this religious form of power. This was a deliberate goal of German theologians Dietrich Bonhoeffer and Jürgen Moltmann and has remained so in many theologies of liberation and almost all feminist theologies. At least initial steps in this direction were made by Karl Barth, Wolfhart Pannenberg, Eberhard Jüngel, and David Tracy, in some process theologies, and by other thinkers and developments. Christological and trinitarian insights and questions were decisive in the efforts to end classical theism (not to be confused with the monotheism of the living God). In addition, insights from the theology of law and from pneumatology, as well as metaphysical, moral, and political arguments forced abstract theism to be called into question. Despite all its difficulties (see Welker, 1999, pp. 1–5), this development has to be supplemented and complemented by an equally serious critique of subjectivist faith.

cannot offer. Faith—at least in the Jewish and Christian traditions—is not only directed to the Creator who sustains his creation. The search for an intensified and deepened knowledge of creation is not sufficient in order to understand the driving energies of faith. Faith goes hand in hand with a deep sense of the endangerment and the self-endangerment of creation and with an awareness that Divine creativity has to supplement the sustaining powers through powers of salvation and redemption. Legal, political, and moral dimensions and their limits in the face of the self-endangerment of human cultures and societies become apparent. Both the Old and the New Testament traditions generated fresh religious insights in the presence of foreign world powers and the inability of the given political and religious traditions to stand up to them. The search for deeper dimensions of the saving and redeeming God guided religious sensitivities.

Yet even the complementarity of Divine sustenance and Divine saving does not explain the full dimensions of faith's orientation as opposed to academic and educational quests for truth. Confronted with the ultimate futility of individual and communal life, including even history and the life of the whole cosmos, faith directs itself toward the Divine elevation and ennoblement of creaturely life. God's rescuing and saving powers at the level of mere natural and historical repair and restitution are not enough. Eschatological questions and hope for the New Creation, the transformation of natural bodies, and a life in realms to which religion, mathematics, and great music possibly bear witness dimly emerge. At that point the clarity and academic controllability of the search for truth becomes questioned. Yet faith cannot abandon these perspectives related to the human search for salvation. It has to search for soteriological and eschatological knowledge.

Subjectivist faith is far from posing these deep questions and challenges. Like Baron von Muenchhausen, who tried to save himself from being swallowed up by the fen by dragging himself out by his own hair, subjectivist faith replaces religious ennoblement by self-referential certainty and a very simple notion of freedom correlated with it. The challenge to investigate demarcations and clashes between faith and knowledge can open one's eyes to the inner logics of current religious decay and to alternatives and opportunities in a complicated cultural setting.

References

Barth, K. (1964). *Kirchliche Dogmatik* [Church dogmatics]. I/1 (8th ed.). Zurich, Switzerland: EVZ-Verlag.

Calvin, J. (1997). *Institutes of the Christian religion*. Edinburgh, Scotland: T&T Clark. (Original work published 1559.)

Henrich, D. (1967). *Fichtes ursprüngliche Einsicht* [Fichte's original insight]. Wissenschaft und Gegenwart No. 34. Frankfurt am Main, Germany: Klostermann.

Henrich, D. (1982). *Selbstverhältnisse* [Self-relations]. Stuttgart, Germany: Reclam.

Kant, I. (2002). *The critique of practical reason* (Werner S. Pluhar, Trans.). Indianapolis, IN: Hackett. (Original work published 1788)

Kierkegaard, S. (1954). *Gesammelte Werke: Bd. 24–25. Die Krankheit zum Tode. Der Hohepriester—der Zöllner—die Sünderin* [Collected Works: Vols. 24–25: The Sickness unto death] (Emanuel Hirsch, Trans). Düsseldorf, Germany: Eugen Diederichs.

Pannenberg, W. (1988). *Systematische Theologie* [Systematic theology] (Vol. 1). Göttingen, Germany: Vandenhoeck & Ruprecht.

Polkinghorne, J. (1994). *The faith of a physicist.* Princeton, NJ: Princeton University Press.

Polkinghorne, J. (2000). *Faith, science and understanding.* London: SPCK.

Polkinghorne, J., & Welker, M. (2001). *Faith in the living God: A dialogue.* London: SPCK.

Polkinghorne, J., & Welker, M. (Eds.). (2000). *The end of the world and the ends of God: Science and theology on eschatology.* Harrisburg, PA: Trinity.

Schleiermacher, F. (1999). *The Christian faith* (Donald Macpherson Baillie, Trans.). Edinburgh, Scotland: T&T Clark. (Original work published 1821–1822.)

Tillich, P. (1951–1963). *Systematic theology* (3 vols.). Chicago, IL: University of Chicago Press.

Welker, M. (1975). *Der Vorgang Autonomie. Philosophische Beiträge zur Einsicht in theologischer Rezeption und Kritik* [The process of autonomy: Philosophical contributions to insight into theological reception and critique]. Neukirchen, Austria: Neukirchener.

Welker, M. (1999). *Creation and reality: Theological and biblical perspectives.* Philadelphia, PA: Fortress.

Welker, M. (2000). Is the autonomous person of European modernity a sustainable model of human personhood? In N. H. Gregersen, W. B. Drees, & U. Görman (Eds.), *The human person in science and theology* (pp. 95–114). Edinburgh, Scotland: T&T Clark.

Welker, M. (2004). Subjectivist "faith" as a religious trap. In W. Schweiker (Ed.), *Having: On property and possession in religious and social life* (pp. 122–137). Grand Rapids, MI: Eerdmans.

Chapter 9
Types of Sacred Space and European Responses to New Religious Movements[1]

Eileen Barker

Throughout history there have been clashes between new religious movements (NRMs) and the societies in which they have arisen. The clashes have occurred not only *within* a particular space (in this chapter I am concerned with contemporary Europe) but also *about* what I shall refer to as sacred space. The chapter starts with an introduction to contemporary new religions and suggests some of the characteristics that might lead to a variety of clashes. It then introduces the concept of sacred space and a typology of different kinds of theological locations of religious identity that could be a further source of clashes between the new religions and those with competing concepts of sacred space. Finally, there is a brief overview of European responses to the movements.

The New Religious Movements (NRMs)

There is no satisfactory definition for the term *new religious movement*. Many of the groups commonly included are not particularly new. ISKCON, the International Society for Krishna Consciousness, insists that it is not an NRM, for it traces its origins back to the 16th-century monk, Chaitanya Mahaprabhu. It is, however, new insofar as it attracted Western converts and developed a new structure when its founder, Prabhupada, arrived in the United States in the 1960s. Other movements are not considered either by themselves or by others as religious, many preferring to describe themselves as spiritual, philosophical, or educational movements, and the Raelians have referred to themselves as an atheistic religion.

The term *new religious movement* has, however, been adopted by scholars in preference to *sect* or *cult* because, although these latter two words have a technical meaning in the sociology of religion, they have come to have a pejorative meaning in popular parlance. Terminology such as *destructive cults* or *sectarian deviations*

[1] I would like to thank the Nuffield Foundation and the Leverhulme Trust for their generous help in funding the research upon which this chapter is based.

P. Meusburger et al. (eds.), *Clashes of Knowledge*.

is commonly used to imply that, unlike "proper" religions, the movements are potentially or actually harmful to individuals or society—or both. The single word *cult* (or *secte* in French) can conjure up the image of a dangerous pseudoreligion possessing satanic overtones, involved in financial rackets and political intrigue, indulging in unnatural sexual practices, abusing its women and children, and using irresistible and irreversible brainwashing techniques to exploit its recruits—and, it may be further assumed, its members are likely to perform all manner of criminal activities, resort to violence and, quite possibly, end up by committing mass suicide.

Although it is undoubtedly true that some new religious movements have been guilty of some of these practices some of the time, it is equally true that some members of old religious movements have indulged in such practices. It is the task of the social scientist to examine each religion and see whether any of these particular sins are part of a particular religion's agenda, rather than lumping together all new religions by means of an emotionally charged label. What immediately becomes apparent when one does investigate different new religions is that one cannot safely generalize about them. They differ from each other according to their beliefs, practices, lifestyle, organization, and leadership; their attitudes toward women, children, finances, and the outside society; and any other characteristic one might imagine.

The movements also differ in the kinds of newness that have been attributed to them. Some have been in existence as mainstream religions in a different geographical space for hundreds or even thousands of years. They are, however, regarded as new religions when they move from East to West—or West to East (Melton, 2004). Sometimes a schismatic group emerges from a mainstream religion, claiming to represent the real, fundamental truths of that tradition. Such groups may be referred to as revivalist movements by some or as new religious movements by others.

One way of approaching NRMs is to take a first-generation membership as the defining characteristic. Although it is still necessary to bear in mind that generalizing about the movements is precarious, such an approach allows one to anticipate certain characteristics that are liable to be present insofar as the movement is new in this sense (Barker, 2004). The very fact that the membership of NRMs consists of converts is likely to have consequences that potentially or actually give rise to tensions and clashes with nonmembers. Firstly, converts are noted for being far more enthusiastic, even fanatic, than people who have been born into their religion. Excited that they have found The Truth, converts frequently feel obliged not only to tell others of their discovery but also to try to persuade them to accept the new beliefs and way of life. This encroachment on the religious sensitivities of their friends and relatives—and, indeed, of complete strangers—can be experienced as irritating, impertinent, or downright offensive.

Secondly, NRMs rarely if ever attract a random sample of the population. In the past they have often appealed to the oppressed, but the current wave of movements that became visible in the West in the 1960s and 1970s appealed disproportionately to well-educated young white adults from the middle classes. This meant that their parents tended to have had different expectations for their children's futures and also were likely to be in a position to make a fuss about their son or daughter's giving up a promising career to follow some guru. It also meant that the young converts were at a stage in their lives when they were unencumbered by financial responsibilities, with the movements having few dependants in the form of children or elderly members to worry about. And it meant that the second-level leadership, while enthusiastic, tended to be inexperienced and, in many ways, immature.

Thirdly, it is common for the founder/leader of an NRM to wield charismatic authority. In other words, the followers believe that the leader has a special quality that gives him (or, occasionally, her) the right to control all aspects of their lives, including where they live and work, whom they may marry, and whether or how they can bring up children. Unlike most leaders of the more established religions, the charismatic leader is not bound by either tradition or rules, and is, thereby, both unaccountable and unpredictable.

Fourthly, new religions very often have a dichotomous worldview. Their knowledge of the world may be uncompromisingly divided into "Godly and satanic"; "right and wrong"; and, drawing a sharp boundary between themselves and the rest of society, "them and us." Drawing such distinctions is not peculiar to the current wave of new religions. Throughout history the members of new religions such as early Buddhism, early Christianity, and hundreds of other fledgling groups have been exhorted by their leaders to detach themselves from their families, friends, and all other nonmembers (see, for example, Luke 14:26 and Matthew 10:35–36). The new religion has carved out its own special space and is prepared to defend its boundaries—sometimes, if necessary, to the death. At the same time, the members of other religions may be equally determined to defend *their* space—their flock, their truths, their knowledge, their ways of life, their vested interests, their cultural heritage, and their future.

But although it is true that a new religion is likely to draw a sharp boundary between itself and the outside society in order to emphasize the difference between "them" and "us" and to protect vulnerable converts from backsliding under the influence of alternative ideas, exactly how the boundaries are drawn differs quite radically from movement to movement.

Varieties of Sacred Space

Different religions have different beliefs about what qualifies and what disqualifies an individual from membership. They have, implicitly or explicitly, a theological definition that locates religious identity within a sacred space. Although these sacred spaces may make reference to physical boundaries, the boundaries delimiting sacred space are conceptual rather than physical. As with all boundaries, the boundaries defining sacred space are more or less clearly defined and more or less permeable. Sometimes they are more permeable in one direction than in another; that is, it may be easier to enter a religion than to leave it, or it may be easier to

Table 9.1 Ideal types of locations of religious identity

Location	Boundary	Primary access
Cosmic	None	Nothing to negotiate
Global	Should be none	Accept
National	Geopolitical	Born into/naturalize
Local	Geospiritual	Birth and/or spiritual commitment to local gods
Biological	DNA inheritance	Born and belong, even if do not believe
Lineage	'Fictive kin'	Initiation
Cultural	Community of believers	Individual negotiability
Individual	Choices	Individual achievement
Inner space	Person	Seek within; limited to self, but there for all
Virtual	Ethereal	Creative negotiability

leave than to enter. The locations of religious identity that I shall describe are what the sociologist Max Weber refers to as ideal types; they are tools that are more or less useful for comparison rather than being more or less true (see Table 9.1). Actual religions may belong to two or more types that overlap in reality, and there are various other possibilities that could have been selected for the typology (for further details, see Barker, 2006). It is hoped, however, that the following depictions will illustrate some of the potentials for "clashes of knowledge" that can be found between different religions.

The Cosmic Location

Perhaps paradoxically for a typology concerned with boundaries, the first type, the cosmic location of religious identity, avers as an integral part of its belief system that there are no boundaries in or to its religion. It asserts that it is the birthright of all humans to belong to a cosmic spirituality—indeed, all humanity *is* part of cosmic spirituality. All one might need to do is to realize one's spiritual nature by "entuning"—getting in touch or resonating with the cosmic vibrations, or, possibly, acquiring the esoteric knowledge of the ultimate Divine. This perspective, which is adopted by many of the groups referred to as the New Age movement,[2] can be seen as threatening by those who have a wish to preserve clearly defined boundaries. It is perceived as undermining clarity, law, and order—and knowledge of what is right and wrong. Among evangelical Christians one can find those who are particularly opposed to the cosmic vision, and there are not a few New Agers who, despite their stated intent to dissolve all barriers, consider evangelical Christians to belong to a "them" category.

[2] This designation has gone out of fashion, but no single term has taken its place in the literature.

The Global Location

At first glance, the global location of religious identity might seem to be similar to the cosmic location. It is, however, quite different. Indeed, evangelical Christians would fall into this type, for they believe that theirs is a universally applicable religion to which all individuals *ought* to convert. It is the task of the evangelical to preach The Word throughout the globe. It is comparatively easy for an individual to be received into this global community by, for example, taking Jesus into his or her heart and accepting Him as their savior. Many, though by no means all, of the new religions consider it to be a central part of their mission to spread their knowledge. Such a conviction may, however, severely test the faith of the believer in situations where proselytizing is outlawed and/or conversion is seen as apostasy and, in extreme circumstances, punishable by death.

The National Location

The national location of religious identity is connected to a geopolitical boundary. Citizenship and membership of the religion are considered to be the same thing. When those within a national religion perceive themselves to be under threat, other religions are seen not as heresy, but as treason. In reality it may well be, as in the case of the Russian Orthodox Church, that there are more members of the national religion living outside the physical boundaries of the state than within it, but such a situation does not prevent the Mother Church from denouncing members of another religion within the country as traitors, even if they have been born and reared in that country and are prepared to fight and die for it. Such condemnation can apply to those who convert either to new religions that originated in foreign lands (such as the Church of Scientology or ISKCON) or to indigenous new religions (such as Vissarion's Church of the Last Testament or the White Brotherhood).

The Local Location

Rather than being geopolitical, local sacred space is geospiritual. This sacred space is tied to the land—to mountains, rivers, and trees. The people who belong to the religion feel they have close ties to the land, but they recognize that there are others who live on the land, have created their own political structures, and have rejected (or are even unaware of) the local gods and/or deities. Pagans whose religious identity is tied to a localized sacred space may have their ethnic or racial roots in the geographical location, as is the case with the majority of worshippers of the Baltic gods; but this is not necessarily so. North America plays host to thousands who worship the Celtic gods. Indeed, according to Matthews (1993), "Celtic ethnicity is not necessarily a prerequisite … [W]e have entered a phase of maturity wherein spiritual lineage transcends blood lineage" (p. 7). This viewpoint has led Bowman (1995) to coin the term "Cardiac Celts" (p. 246) to describe those who feel in their hearts that they are Celts. The point here is that the spiritual or sacred space is connected to a particular locality through spiritual rather than biological or physical proximity. The majority of pagans are not overly concerned with physical boundaries, but there are some pagans in parts of the former Soviet Union, Scandinavia, and elsewhere whose beliefs are aggressively laced with a strident racism and who express unambiguous hostility toward those whom they consider ethnic interlopers, including Christians, whom they consider to have usurped their sacred space. Another example of a religion that stresses the local location is "Shrine Shinto" (as opposed to "State Shinto," which belongs to the previous, national type), yet Brazil plays host to thousands of Japanese who worship *kami* whose primary location is in the Land of the Rising Sun.

The Biological Lineage

For members of the biological type of theological location, their religious identity would seem to be located in their DNA, whether or not they accept the beliefs of the religion or practice its rituals. In the Jewish tradition, membership is passed through the mother's line; in the Zoroastrian tradition, it is passed through the father's. Clearly, a biological lineage is a sacred location that would be particularly difficult to permeate. Nonetheless, it is possible to convert to Judaism if one persists, and there is currently a debate within Zoroastrianism as to whether the children of women who marry "out" can be accepted as Parsis.

Converting to a new (or old) religion by someone born into a biologically transmitted religion can be seen as denying one's true nature and undermining the continuance of the race or ethnic group. But clashes of knowledge can be particularly acute when the convert joins a group that, despite acknowledging the biological heritage, embraces tenets that seem to contradict the mainstream understanding of the biological religion, as when a Jew joins a messianic religion such as Jews for Jesus.

Given that new religions are dependent on the convert, it might seem, *prima facie*, that they could not claim a biological location for the religious identity of their membership. There are, however, new religions that, with the arrival of a second generation, have established a sacred lineage that gives a special place to those who have been born into the movement. The Unification Church, for example, believes that the children of couples who have been Blessed by Sun Myung Moon will be free of "fallen nature" and thus able to lay a foundation for reestablishing God's Kingdom of Heaven on earth. The movement makes a sharp distinction between these "Blessed children" and others (known as "Jacob's children"), whose parents joined the movement after they were born and who are, therefore, still the inheritors of fallen nature.

The Religious Lineage

Sacred identity that is conferred through a religious rather than a biological lineage is frequently found in Eastern religions that have a tradition of devotees inheriting their religious credentials by following a guru, the gurus having, in turn, acquired their credentials from their gurus. His Divine Grace A. C. Bhaktivedanta Swami Prabhupada, for example, who had received his knowledge from a guru who traced his lineage back to Lord Chaitanya, initiated a large number of disciples when he came to the West. Since his death, however, disagreements over successions and arguments about who is really qualified to transmit the lineage have been rife both within ISKCON itself and among a number of schismatic groups that make counterclaims about the true transmission of the lineage and the authenticity of subsequent gurus.

Religious lineages are also found in a variety of African, Caribbean, and Native American religions where an apprenticeship is served with a Sangoma, Shaman, or Master. Traditionally, considerable time and concentration has had to be invested in the transmission process, but several new religions have been offering an instant enlightenment that has angered those who consider that their traditions are being undermined by overly easy access to (and departures from) sacred lineages.

The Cultural Location

Allocation to a cultural sacred space is apt to be through birth into a religious group that is associated with a particular historical heritage. The child's identity within this sacred space is usually confirmed with some ritual of public affirmation, which can occur, as with infant baptism, without the individual's considered consent or even knowledge. It may be relatively easy to join or leave a cultural religion, although (as with the biological type of religious identity) it is possible that people born into it are considered more genuine members than converts, who may be regarded with a certain degree of suspicion. It is also possible that those "born-intos" who have not formally accepted membership or do not attend the religion's services will nonetheless consider themselves members so far as their religious identity is concerned. I found that there was actually a higher proportion of Britons who felt they belonged to the Church of England than were actual members of the Church.[3]

[3] This was part of a pan-European study into Religious and Moral Pluralism (RAMP), conducted in 1998, that involved a representative sample of 1,466 British residents aged 18 or over.

Religions that allocate membership to a cultural space may share some characteristics with the national religions, but geographical or political boundaries are not so fiercely defended and the religious culture can move more easily into what might be viewed as different cultural space.[4] Almost by definition, NRMs have not had time to create a cultural religion, but converts *from* such religions can provoke tensions not dissimilar to, though usually not as strident as, those that result from conversion from a national religion. Just as with conversion from the biological type, there is a feeling that converts have rejected part of their core identity—they are no longer viewed as the same person.

The Individual Location

Religions that are associated with the individual location of religious identity consider personal choice and commitment to be the prime, if not the only, way in which an individual can be counted as a member. It differs from the universal type in that there is no claim that everyone *ought* to be a member; it is entirely up to each person to decide. Both entry and exit is a relatively simple matter with little in the way of restriction in either direction. The boundary is not tied to social, cultural, political, or geographical locations, only to where one's conscience lies. Several new religions might claim to fall into this category, and, being religions of personal choice, can be seen as a threat by traditional religions that wish to draw on secular locations to reinforce and be reinforced by religious commitment. There are, however, also older religions that fit the mould, the Society of Friends (or Quakers) being but one example.

The Location of Inner Space

Whereas religious identity for the previous type is located in a group with a particular set of beliefs and practices with which individuals have chosen to associate themselves, one's sacred identity in inner space is located not so much in the group as within oneself. The individual is expected to go within himself or herself to seek "the god within." Heelas (1982) has referred to some new religions of this kind as "self religions," but there are several varieties of both new and traditional groups that believe in and celebrate the inner sacred space that defines their religious identity.

[4] The Anglican Communion presents an interesting case at the present time, with African congregations insisting on a more literal interpretation of Anglican beliefs than do most English believers. Here the clashes of knowledge are occurring within the boundaries of a religious community rather than at or across the boundary, with those at the geographical periphery claiming spiritual centrality in their endeavors to missionize those at the geographical center.

Curiously enough, this type, which might be seen as having little association with physical, biological, or social space, is in several ways closely related to the type that is associated with the widest expanse of space—the cosmic location. Both undermine the embeddedness of the boundaries of other types, not so much by attacking them at their frontiers as by ignoring them or denying their salience. In other words, although the primary focus of the two types differs, they can merge into each other as part of what is sometimes referred to as "the new spirituality."

The Virtual Location

The final type of location for religious identity is the only one that is genuinely new to contemporary society. Virtual sacred space exists through the worldwide web. Identities can be created and dissolved without any of the normal restrictions of gender, age, nationality, culture, class, geography, or DNA.[5] Individuals are free to create their own identity through the medium of the Internet. They can create, join, adjust, or leave virtual religions, spiritual communities, groups, or alternative realities, restricted only by their ability to make the appropriate clicks on a mouse or keyboard in the privacy of their rooms or in an Internet café. Those boundaries that are created in the ether do not challenge traditional religions directly, but they do offer alternative identities for those who retreat into them, creating an exodus from the "real life" religions of the world.

Much more could be written about sacred space and religious allocations of identity. It is hoped, however, that enough has been said to indicate how there can be clashes of knowledge over where true religion lies, who is entitled or expected to belong or not belong, and how there can be asymmetrical degrees of negotiability over the crossing of boundaries that define the locations of sacred space. A critical point is that all the boundaries *are* negotiable, though some might appear not to be, and this negotiability suggests that there could be a constant dynamic related to the kinds of clashes that can occur between the various types outlined above.

Reactions to the NRMs

A suspicion and fear of the new religions has been fanned by a number of tragic episodes involving deaths and criminal activity associated with a few of the movements. Among the first of these incidents to hit the headlines in the West were the

[5] Of course, it is unlikely that any human being is unaffected and uninfluenced by all these "worldly" variables, however much they might try to transcend them. The point is, however, that the potential for negotiability of religious identity in virtual sacred space is considerably greater at the conscious if not the unconscious level than is likely to be the case in any of the other sacred locations.

Manson Family murders in 1969, when Sharon Tate and six others were killed by followers of Charles Manson in California. The United States also experienced the Symbionese Liberation Army's kidnapping of Patty Hearst and her subsequent involvement in an armed robbery in 1974. Another catastrophe involving a U.S.-related NRM was the killing of U.S. Congressman Leo Ryan and his companions, followed by the suicides and murders of over 900 members of the People's Temple in Jonestown, Guyana, in 1978. And in 1993 there was the fiasco outside Waco, Texas, when the FBI stormed the Mount Carmel compound after a 51-day siege, resulting in the death of over 80 Branch Davidians, including 24 children and their leader, David Koresh.

The first such tragedy to hit Europe was the suicides and murders of members of the Solar Temple in Switzerland and France between 1994 and 1997. Then the whole world became increasingly concerned when members of Aum Shinrikyo deposited sarin gas in the Tokyo underground in 1995. This atrocity, which resulted in 12 deaths and over 1,000 people being injured, was the first occasion on which a "weapon of mass destruction" had been used and in which the target had been not members of the movements themselves or people known to them, but innocent citizens going about their daily lives in what could be described as neutral space.[6]

Just as it is hazardous to generalize about the new religions, so is it folly to generalize about reactions to them. Even when the discussion is confined to Europe, the briefest of surveys indicates that the diversity is considerable. Richardson (2004) has drawn a useful preliminary distinction between two types of position. On the one hand, there are countries (such as the United States, Scandinavia, the Netherlands, and the United Kingdom) that favor "self-help remedies," where individual circumstances are dealt with privately or cases are brought to the courts to rectify perceived violations of some general value or law. On the other hand, there are countries (such as China, Russia, Belgium, and France) where it is the state that assumes responsibility to protect its citizens from both the actual and the potential harm of NRMs. This type of state involvement might include "public information campaigns" to warn the public about the "cult menace"; or it might entail discriminatory regulation of the behavior of minority religions, irrespective of whether any actual crime has been committed. Of course, as with all typologies, Richardson's distinction is useful only as an initial orientation for comparative purposes. The actual situation within each country, and in Europe as a whole, is continuously altering because of the transformations brought about by, for example, social and geographical mobility, the mass media and the Internet, and, not least, radical developments within the NRMs themselves.[7]

---

[6] It is true that the Kasumagaseki underground station which was targeted was near the National Police Agency's headquarters, but it was commuters in general rather than the police in particular who were the anticipated and actual victims.

[7] A key characteristic of new religions is that they are almost bound to change far more rapidly than older, more established religions, if only because of demographic changes such as the aging of converts, the arrival of second and subsequent generations and the eventual death of charismatic founders (Barker, 2004).

One of the first European governmental reactions to post-war NRMs was in 1982, when the then-Prime Minister of France, Pierre Mauroy, charged Alain Vivien (a Deputy in the National Assembly) with writing a report on the movements. Although the report (Vivien, 1985) recommended various measures to control potential dangers posed by the movements, nothing much was done. Over the next decade, further reports were commissioned by various governments as well as by both the European Parliament (Berger, 1997; Cottrell, 1984) and the Council of Europe (Hunt, 1991; Nastase, 1998). Some of the reports, such as those of The Netherlands (Witteveen, 1984), Germany (Schätzle, 1998), and Sweden (Ingvardsson et al., 1998), concluded that, although some measures might have to be introduced to provide further information about the movements, the existing law seemed, on the whole, to be sufficient to deal with them. Others reports, such as those of Russia (Kulikov, 1996), France (Gest & Guyard, 1995), and Belgium (Duquesne & Willems, 1997), concluded that special legislation should be introduced to control the movements.[8]

The French and Belgian reports were among those that had the greatest effect. Firstly, they both contained a list of "harmful sects," which included such groups as, in the Belgian case, the Quakers, the Mormons, and the YWCA (Young Women's Christian Association). Although neither government formally accepted the lists, their existence has led to numerous complaints that they have given semi-official organizations and the general public permission to discriminate against the named movements in such matters as renting halls, buying property, educating their children, and obtaining employment. At the official level, the Belgian government set up a cult-watching organization, CIAOSN,[9] and the French government instigated MILS,[10] which, in 2000, proposed a bill for the "Prevention and Repression of Sect Movements" that was passed in an amended version the following year. The law's provisions included the extension of legal responsibility from individuals to organizations, which could be dissolved if they or their executives were found guilty of a serious crime. Although the law does not specifically mention sects, or even religion, one of its co-architects, Catherine Picard, is reported to have said "We need to give judges repressive tools. The law is a response to the evolution of society and the growing importance that sects have in it" (Bosco, 2001).

Apart from being signatories to statements such as the United Nations Declaration of Human Rights and the European Convention on Human Rights, most European countries have constitutions declaring that the state endorses religious freedom for all citizens, even if it gives a special status to one or more favored

[8] The different approaches of the Reports and their recommendations are discussed in some detail in Richardson and Introvigne (2001).

[9] *Le Centre d'Information et d'Avis sur les Organisations Sectaires Nuisibles* (Information and Advice Center on Harmful Sectarian Organizations).

[10] *Mission Interministérielle de Lutte contre les Sectes* (Interministerial Mission in the Fight Against Cults), which was later replaced by MIVILUDES (*Mission Interministérielle de Vigilance et de Lutte Contre les Dérives Sectaires*).

religions. Not infrequently, officials of these favored religions (notably Orthodox Churches in Greece and Eastern Europe, and to some extent the Catholic Church in Poland) have played a significant role both in urging the enactment of laws that would restrict the practices of minority religions and in sanctioning discriminatory treatment of the movements, especially at the local or regional level.

Many, though not all, European countries have now introduced laws related to the registration of religions. The criteria for registration vary, but commonly they refer to the number of members and the length of time the group has existed in the country, both of which can disadvantage newer religions. Sometimes (as in Lithuania) there are different tiers of recognition, the newest religions receiving the fewest privileges. Slovakia requires the endorsement of at least 20,000 adult residents before a religion can be registered, a number that few new religions can hope to have achieved. Indeed, only the Jehovah's Witnesses and the Church of Jesus Christ of Latter-day Saints (the Mormons), both founded in the 19th century, have succeeded in obtaining sufficient numbers. This is no accident. The Slovakian Ministry of the Interior's *Statement of Reasons* for a proposed tightening of the relevant law declares: "This provision should restrict the potential for abuse by 'religious' groups of dubious origin seeking the legal status and benefits that registered churches and religious organizations are afforded in Slovakia" (Slovakian Ministry of the Interior, 2007).

It should, however, be noted, firstly, that all religions, whether or not they are registered, are allowed to operate freely in Slovakia so long as their members obey the law and, secondly, that registration leads to benefits, such as state subsidies and the right to teach in public schools. In other societies lack of registration can be a more serious disadvantage. Under Article 14 of the Russian 1996 Law, for example, religions that do not succeed in getting registered may be "liquidated" (Uzzell, 1998). Jehovah's Witnesses, since the liquidation of their community in Moscow (Golovinsky Intermunicipal District Court, 26 March 2004), have been denied the right to rent public halls for meetings, have been arrested for discussing the Bible in private homes, and, on occasion, have been physically assaulted, jailed, or both (Office of Public Information for Jehovah's Witnesses, 2005).

In Germany the new religions are, on the whole, allowed to practice without interference. Two exceptions are Scientology and Hizb-ut-Tahrir. The Scientologists have long been the chief target of German cult-watching groups (Kniola, 1997), particularly in Hamburg and in Bavaria, where a "sect filter" requires public employees, including university teachers, to sign a form attesting that they are not members of Scientology or other extremist groups. Hizb-ut-Tahrir, an Islamic NRM founded in 1953, has as its main objective the establishment of the Caliphate throughout (at least) the Muslim world. Its activities are banned in Germany (its members cannot book a hall or give public talks), but membership of the movement is permitted. During his period of office as the British Prime Minister, Tony Blair stated that he would like an outright ban on Hizb-ut-Tahrir in Britain, but this proposal drew opposition from several quarters, including, it was reported, the Home Office and the security services (BBC News 24, 2006). At the time of writing (2007), the movement is still operating freely. However, there are two groups on the

U.K.'s list of proscribed terrorist organizations: Al-Ghurabaa and the Saved (or Saviour) Sect, both of which appeared after the disbanding of Al-Muhajiroun, apparently as its successors (Home Office, n.d.).

Of course, reactions to NRMs are not confined to antisect laws. Perfectly neutral laws may be applied in a discriminatory fashion, and courts may disallow evidence that would be of benefit to an unpopular religion, while allowing evidence that would be prejudicial and considered inadmissible in other circumstances (Barker, 1987). Such decisions were made on a number of occasions when "deprogramming" (the illegal kidnapping and holding of members of NRMs against their will) was a relatively common method for "rescuing victims" from the movements (Barker, 1989). To cite but a few further examples: police are reported to have stood by while radical members of the Georgian Orthodox Church beat up Jehovah's Witnesses (Amnesty International, 2001). Scores of Jehovah's Witnesses have been imprisoned in Armenia because of their stand as conscientious objectors, despite the state's commitment to the Council of Europe to provide for civilian alternative service. Several leaders of movements have been denied entry to European countries. Louis Farrakhan of the Nation of Islam and Sun Myung Moon have both been denied visas by the United Kingdom,[11] and Raël, founder of the Raelian movement, has been denied residency in the Swiss Canton of Valais "for fear of endangering public morals" ("Cult leader," 2007).

Other institutions and members of the general public can welcome, tolerate, revile, denounce, or attack the new kid on the block with just as much, if not more, effect than the state. Plenty has been written about the media and the Internet and how they can influence the reception and perception of the movements (Beckford, 1999; Hadden & Cowan, 2000). There are also various types of cult-watching groups that draw different images of the new religions, selecting what they consider to be salient information according to their own particular interests (Barker, 2002). The net result is frequently the dissemination of a conventional wisdom that generalizes about the new religions, labeling them as destructive cults or dangerous sects by circulating what have been called "atrocity tales" through personal networks and in the popular media.

None of this is to suggest that NRMs do not deserve some of the bad press that they receive. But it is worth pointing out that there are hundreds, if not thousands, of NRMs in Europe and that the vast majority of them have not indulged in any criminal behavior.[12] It is also worth recognizing that it is relatively easy to employ double standards when judging the movements and to forget that members of older religions and, indeed, of no religion, have also indulged in criminal and antisocial actions.

---

[11] More recently, Moon has been allowed into Britain on a short-stay visa.

[12] The exact number depends on precisely how an NRM is defined and whether one places together small groups, such as individual covens, under one label (Wicca). At Inform (www.Inform.ac), a government-supported charity I founded to provide information about the movements, there is information about 1,001 identifiable groups currently active in the United Kingdom.

There has been some vocal criticism from Human Rights groups, such as the Belgian-based Human Rights Without Frontiers (http://www.hrwf.net), Forum 18 (http://www.forum18.org), and the International Helsinki Federation for Human Rights (http://www.ihf-hr.org), academics (e.g., Introvigne & Melton, 1996; Richardson, 2004), and sections of the media (see, for example, the collection of articles on the website of WorldWide Religious news, http://www.wwrn.org) about the treatment of NRMs in parts of both western and eastern Europe. The Organization for Security and Co-operation in Europe (OSCE, http://www.osce.org) has hosted numerous meetings at which complaints have been raised about discrimination. But particularly irksome to the governments concerned have been entries in the U.S. State Department's annual Human Rights Report, and a series of Congressional hearings at which, for example, the U.S. Ambassador-at-Large for International Religious Freedom objected to the treatment of minority religions in Germany, Austria, France, and Belgium (*Discrimination*, 2000).

The movements themselves have reacted in a variety of ways to perceived and actual discrimination. Some, such as the Children of God in the 1970s and 1980s, have gone underground; others have responded in the courts and by other, sometimes aggressive and/or dubious, means. Both the Church of Scientology and the Jehovah's Witnesses have resorted to the law on numerous occasions and, in so doing, have tested the boundaries of publicly acceptable behavior by both members and nonmembers. The Witnesses have won a number of cases in the European Court of Human Rights (ECHR), including the oft-cited Kokkinakis case concerning antiproselytizing laws in Greece (*Kokkinakis v. Greece*, 1993). At the time of writing, the Witnesses are awaiting outcomes of their ECHR appeal cases against the Moscow court's decision to liquidate them and against the French tax department's decision to levy a 60 percent tax on donations the movement has received (Office of Public Information for Jehovah's Witnesses, n.d.).

Other things being equal (which, of course, they seldom are), minority religions are less likely to be subjected to discriminatory treatment in any particular country insofar as (a) there exists a strong tradition of and support for democratic values; (b) the country is a signatory of International Human Rights Declarations; (c) the Constitution and laws of the country guarantee freedom of religious beliefs and practices that remain within the law; (d) the country has a history of religious pluralism; (e) the media are open and subject to complaint procedures; (f) relatively easy access to legal appeals offers the possibility of redress in the face of unlawful discrimination; (g) the country's government and justice system are impartial and not corrupt; (h) human rights and civil-liberty organizations have a voice that can be heard; (i) accurate information is available about the movements' beliefs and practices; (j) the state and/or its citizens do not perceive themselves (rightly or wrongly) to be under some sort of economic, political, or military threat or in danger of losing their national identity; (k) the rate of immigration from an unfamiliar culture is not high; and (l) the behavior of the NRMs is open and in accordance with the law of the land.

Again, other things being equal, one might also argue that societies in which the dominant religions define the sacred space of their religious identity in terms of a national location are particularly likely to object to the appearance of new religions. At the other extreme, although the Internet can certainly be used to spread hatred of other religions, religions that are themselves located in virtual space are relatively unlikely to worry about their neighbors' religion so long as it does not directly interfere with their own beliefs. It might, moreover, be suggested that the cosmic and the inner locations of religious identity are less likely to give rise to clashes than the global or cultural locations are. But, as stated earlier, boundaries are negotiable, and there are many other factors that always need to be taken into account. Just as it has long been acknowledged that the occupation of physical space can form the basis for clashes of knowledge between many different types of religious and spiritual positions, perhaps it is also necessary to recognize more clearly the potential and actual role of sacred space in such clashes.

References

Amnesty International. (2001). Georgia: Appeal Cases, 19 September 2001(1); AI Index: EUR 56/011/2001.

Barker, E. (1987). The British right to discriminate. In T. Robbins & R. Robertson (Eds.). *Church-State relations: Tensions and transitions* (pp. 269–280). New Brunswick, NJ, & London: Transaction.

Barker, E. (1989). *New religious movements: A practical introduction*. London: HMSO.

Barker, E. (2002). Watching for violence: A comparative analysis of the roles of five cult-watching groups. In D. G. Bromley & J. G. Melton (Eds.), *Cults, religion and violence* (pp. 123–148). Cambridge, England: Cambridge University Press.

Barker, E. (2004). What are we studying? A sociological case for keeping the "nova." *Nova Religio, 8*(1), 88–102.

Barker, E. (2006). We've got to draw the line somewhere: An exploration of boundaries that define locations of religious identity. *Social Compass, 53*(2), 201–213.

BBC News 24. (2006, November 19). Blair bid to ban group 'opposed'. Retrieved May 16, 2007, from http://news.bbc.co.uk/1/hi/uk/6162690.stm

Beckford, J. A. (1999). The mass media and new religious movements. In B. R. Wilson & J. Cresswell (Eds.), *New religious movements: Challenge and response* (pp. 103–119). London: Routledge.

Berger, M. (1997). *Draft report on cults in the European Union*. Strasbourg, France: European Parliament Committee on Civil Liberties and Internal Affairs.

Bosco, J. (2001, July 10). China's French connection. *Washington Times* [Editorial]. Retrieved May 16, 2007, from http://www.cesnur.org/2001/falun_july03.htm#Anchor-11481

Bowman, M. (1995). Cardiac Celts: Images of the Celts in paganism. In G. Harvey & C. Hardman (Eds.), *Paganism today: Wiccans, Druids, the Goddess and ancient earth traditions for the twenty-first century* (pp. 242–251). London: Thorsons.

Cottrell, R. (1984). *The activity of certain new religions within the European Community*. Strasbourg, France: European Parliament Working Paper of the Committee on Youth, Culture, Education, Information and Sport.

Cult leader. (2007, February 19). Cult leader Rael denied residence in Switzerland. *Breitbart.com* Retrieved May 16, 2007, from http://www.breitbart.com/article.php?id=070219183652.zb2e8qnp&show_article=1

Discrimination. (2000, June 14). *Discrimination on the basis of religion and belief in Western Europe: Hearings before the House International Relations Committee*(testimony of Robert A. Seiple). Retrieved May 16, 2007 from http://www.cesnur.org/testi/EUintolerance_02.htm

Duquesne, A., & Willems, L. (1997). *Enquete Parlementaire visant à élaborer une politique en vue de lutter contre les pratiques illégales des sectes et le danger qu'elles représentent pour la société et pour les personnes, particulièrement les mineurs d'âge* [Parliamentary enquiry aiming to develop a policy to combat the illegal practices of sects and the danger that they pose to society and individuals, particularly minors]. Brussels: Belgian House of Representatives.

Gest, A., & Guyard, J. (1995). *Les Sectes en France* [Sects in France]. Paris: Assemblée Nationale.

Hadden, J. K., & Cowan, D. E. (Eds.). (2000). *Religion on the Internet: Research prospects and promises*. Amsterdam & London: JAI.

Heelas, P. (1982). Californian self religions and socializing the subjective. In E. Barker (Ed.), *New religious movements: A perspective for understanding society* (pp. 69–85). Lewiston, NY: Edwin Mellen.

Home Office. (n.d.). List of proscribed terrorist groups. Retrieved May 16, 2007, from http://www.homeoffice.gov.uk/security/terrorism-and-the-law/terrorism-act/proscribed-groups

Hunt, J. (1991). *Report on sects and new religious movements*. Strasbourg, France: Council of Europe Parliamentary Assembly Committee on Legal Affairs and Human Rights.

Ingvardsson, M., Wallbom, S., & Grip, L. (1998). *I God Tro: Samhället och nyandligheten* [In good faith: Society and the new religious movements]. Stockholm: Statens offentliga utredningar, Socialdepartementet.

Introvigne, M., & Melton, J. G. (Eds.). (1996). *Pour en Finir avec les Sectes: Le débat sur le rapport de la commission parlementaire* [The last word on cults? The debate on the French Parliamentary Report]. Turin, Paris: CESNUR.

Kniola, F.-J. (1997). *Zur Frage der Beobachtung der Scientology-Organisation durch die Verfass ungsschutzbehörden* [On the surveillance of the Scientology organization by the authorities for the protection of the Constitution]. Düsseldorf, Germany: Innenministerium des Landes Nordrhein-Westfalen.

Kokkinakis v. Greece.(1993). (Application No. 260-A) Judgment of 25 May 1993, Ser. A.; 17 EHRR 397.

Kulikov, A. (1996). *Inquiry on the activities of certain foreign religious organizations, gathered from materials of the MVD, FSB, Ministry of Health, Ministry of Welfare, and General Attorney's Office*. Moscow: Ministry of Internal Affairs of the Russian Federation.

Matthews, C. (1993). A Celtic quest. In V. Barnett, R. Howarth, & P. WIlliams (Eds.), *Exploring journeys* (pp. 1–15). London: Shap Working Party on World Religions in Education.

Melton, J. G. (2004). Toward a definition of "new religion." *Nova Religio 8*(1), 73–87.

Nastase, A. (1998). *Illegal activities of sects*. Strasbourg, France: The Council of Europe: Religious Intolerance and Discrimination Section.

Office of Public Information for Jehovah's Witnesses. (2005, May). Effects of Moscow ban on Jehovah's Witnesses. Retrieved May 16, 2007 from http://www.jw-media.org/region/europe/russia/english/moscow/rus_e0506.pdf

Office of Public Information for Jehovah's Witnesses. (n.d.). Europe. Retrieved May 16, from 2007 from http://www.jw-media.org/newsroom/index.htm?content=europe.htm

Richardson, J. T. (Ed.). (2004). *Regulating religion: Case studies from around the globe*. New York & Dordrecht, The Netherlands: Kluwer/Plenum.

Richardson, J. T., & Introvigne, M. (2001). "Brainwashing" theories in European parliamentary and administrative reports on "cults" and "sects." *Journal for the Scientific Study of Religion, 40*(2), 143–168.

Schätzle, O. (1998). *New religious and ideological communities and psychogroups in the Federal Republic of Germany: Final report of the Enquete commission on "So-called Sects and Psychogroups."* Bonn, Germany: Deutscher Bundestag.

Slovakian Ministry of the Interior. (2007). Proposed Amendment 97 to Act. No. 192/1992 on the Registration of Churches and Religious Organizations (2007).

Uzzell, L. (Trans.). (1998). The 1997 Russian Federation federal law on freedom of conscience and on religious associations. *Emory International Law Review, 12*(1), 657–680.

Vivien, A. (1985). *Les Sectes en France: Expression de la Liberté morale ou Facteurs de Manipulations?* [Sects in France: Expression of moral liberty or factors of manipulation?]. Paris: La Documentation Française.

Witteveen, T. (1984). *Overheid En Nieuwe Religeuze Bewegingen* [Overview of new religious movements]. The Hague, The Netherlands: Tweede Kamer.

Chapter 10
When Faiths Collide: The Case of Fundamentalism

Roger W. Stump

Since the late 19th century, fundamentalism has arisen as a major force in the contesting of knowledge within a variety of religious traditions, including most major world religions (Marty & Appleby, 1991–1995). Indeed, the contesting of knowledge represents a defining characteristic of fundamentalism as a cultural, social, and political phenomenon. Clashes of knowledge involving fundamentalists have typically centered on issues of religious authority and authenticity but have involved secular concerns as well, such as the role of religion in society or the relationship between religion and science. As a result, the emergence of clashes of knowledge involving fundamentalists has had far-reaching effects in diverse geographical settings.

This chapter articulates the sources and structures of fundamentalist clashes with other systems of knowledge and the types of meanings that fundamentalists most frequently contest. The discussion focuses in particular on the inherent selectivity of fundamentalist concerns in clashes over religious knowledge and on the role of context in shaping fundamentalist motivations, goals, and actions in such conflicts (Stump, 2000). In addressing these issues, the chapter begins with a general consideration of the nature of religious knowledge, including the transformation of religious knowledge by innovative movements within specific contexts and the distinctiveness of fundamentalism as one such form of innovation. Within this framework, the argument then addresses the selective and contextual nature of fundamentalist concerns and the relationships of those concerns to perceived threats to fundamentalist conceptions of religious knowledge. The discussion concludes with an assessment of how fundamentalists in different contexts engage in clashes of knowledge linked to perceived threats, pursuing goals associated with their religious certainties.

Context and Religious Knowledge

The clashes of knowledge involving fundamentalism have arisen in relation to basic elements of religions as cultural systems. Scholars have defined religions as cultural systems in various ways (Geertz, 1973, pp. 87–125; Pals, 1996, pp. 269–270;

P. Meusburger et al. (eds.), *Clashes of Knowledge*.

Smith, 1996). The definition employed in this chapter identifies religion as a compelling, integrated system of beliefs and practices enacted and reproduced by adherents, a system that relates human life to the existence of a superhuman being or beings (Stark & Bainbridge, 1987, pp. 39–40). The structure of religious knowledge within such systems is, in turn, organized around two formative sources of meaning: a worldview and an ethos (Geertz, 1973, pp. 126–141). The worldview of a religion encompasses an understanding of the nature of reality, including conceptions of causation and agency within the cosmos and their relation to superhuman forces. A religion's ethos relates human thought and behavior to the reality of the worldview, defining basic norms, structures of daily life, and pervasive emotional patterns. A religion's worldview in essence shapes adherents' knowledge of truth and faith, its ethos knowledge of legitimacy and proper action. Together, these sources of meaning express powerful meanings concerning the cosmic order and the essence of human existence, providing an integrated foundation from which adherents understand the world around them and their place within it.

Adherents understand their worldview and ethos to be primary sources of religious certainty, applicable in all times and places as expressions of fixed, eternal meanings. Indeed, certain elements of a religion's worldview and ethos may be relatively invariable across different populations of adherents. Acceptance of Muhammad as God's final prophet is a common element of the worldviews of Muslims, for example, just as the ethos of personal salvation through the atonement of Christ is shared by diverse groups of Christians. Such unifying certainties represent the core of a religious tradition and are fairly resistant to change. As cultural systems, however, religions are repeatedly transformed as their adherents reproduce them in particular contexts. Thus, while adherents believe that their religion is based on immutable truths, that view does not preclude the development of substantial local variations in specific beliefs and practices (Charlesworth, 1997, pp. 81–104).

Within a larger religious tradition, then, religious diversity often develops across space and time, engendering variety in belief through the relationships between religious knowledge and the contexts of adherents' lives. From this perspective, it is useful to distinguish between religious traditions and specific religious systems. A religious tradition comprises a general set of religious givens widely accepted by adherents in diverse settings. Over time, however, a religious tradition may take on varied expressions as spatially dispersed adherents transform it into local religious systems. Using this terminology, Hinduism, Buddhism, Judaism, Christianity, and Islam represent religious traditions whose core bodies of knowledge underlie a variety of distinct religious systems derived from them. The adherents of a specific religious system reproduce a larger tradition's common body of knowledge in their own way, according to their own experiences. The resulting contextuality of religious systems represents one of their crucial attributes as cultural phenomena.

Within Christianity, for example, the tradition's common body of religious knowledge has undergone a long history of transformation by adherents in different geographical contexts. Some transformations have developed on wider scales, through the division of Christianity into Roman Catholicism, Eastern Orthodoxy, and Protestantism, and through the subdivision of the latter into distinct confessions. Within each of these major branches, however, folk and popular interpretations have produced many versions of Christianity on more local scales, some through the creation of new church structures and others through the adaptation of existing structures to local contexts (see Badone, 1990; Conkin, 1997; Nolan & Nolan, 1992). In the process, adherents of these contextual versions of Christianity have reshaped their larger tradition to accord with their own lived experience.

Local transformations of religious knowledge take many forms, of course, from unreflexive, incremental change to explicit expressions of schism and sectarianism. In commonplace patterns of incremental change, the transformation of religious knowledge occurs gradually. Associated changes in worldview and ethos remain relatively minor and do not cause significant controversy; and adherents may have little sense of remaking their religious tradition as they adapt it to their particular context. Folk religions often follow this pattern, examples being the worship of local saints, the formation of local pilgrimages, or the addition of local accretions to established rituals. The syncretistic merger of a larger tradition with other local religious influences represents a more distinctive form of the same pattern.

Other contextual transformations of a religious tradition, including fundamentalism, tend to be both more reflexive and more controversial than those that develop incrementally and unreflexively. These more discordant movements may support clashes in religious knowledge on a broad scale, for example, by generating a major schism within an existing tradition. The concept of schism generally applies to a division between prominent groups within a religious tradition, both of which claim to be the legitimate bearers of the tradition's knowledge system. Schismatic movements typically arise through substantial reinterpretations of a tradition's worldview, which subsequently generate equally substantial changes in its ethos. The corresponding recasting of basic concepts of authority and authenticity may, in turn, provoke significant clashes over the nature of orthodoxy between schismatic and conventional adherents. Such conflicts have developed frequently in the world's major religions, most conspicuously through the emergence of separate doctrinal branches, such as Roman Catholicism, Eastern Orthodoxy, and Protestantism in Christianity, the Sunni and Shi'a branches within Islam, and the Mahayana and Theravada branches within Buddhism.

On narrower scales, elements of schismatic patterns of change appear in the creation of smaller sects and cults. Sect and cult formation generally involves more idiosyncratic and extensive departures from the worldviews and ethoses of antecedent traditions. Sect formation commonly arises from the founders' concerns with authentic interpretation of a tradition's worldview as well as with their emphasis on particular beliefs perceived to be especially crucial, such as specific scriptural passages or ritual practices. Sect adherents also stress the importance of separating from other followers of their larger tradition through the creation of a distinct religious organization (Stark & Bainbridge, 1987, p. 124). Rather than challenging orthodoxy, sects attempt to redefine it in their own terms, although this change may generate significant conflict with others in their tradition. Cults resemble sects in their emphasis on organizing separately from existing religious bodies but are much more idiosyncratic in their interpretations of religious knowledge. Typically, they replace many orthodox beliefs and practices with new and often radical religious conceptions (Stark & Bainbridge, 1987, p. 157). The rejection of a tradition's prevailing worldview thus tends to be more far-reaching in cults than in sects, often supporting adherence to an original ethos.

In comparison to other reflexive patterns of religious change, fundamentalism represents a rather different form of innovation, possessing a number of distinctive characteristics. Most important, fundamentalism has developed as an explicitly modern phenomenon, emerging in response to modern patterns of social change and employing modern strategies and methods to pursue specific goals. In addition, fundamentalist movements develop around intensely reflexive narratives of identity and legitimacy that articulate dialectical relationships to other cultural influences within a given context. To a greater extent than other forms of religious transformation, moreover, fundamentalist movements are less concerned within advancing new expressions of religious knowledge than with fostering authentic and authoritative understandings of existing religious certainties. Unlike schismatic groups, sects, and cults, therefore, fundamentalist groups do not necessarily seek institutional separation from existing religious bodies and may instead try to promote their ideas within existing institutions. Fundamentalists have also typically interpreted their ethos as requiring engagement with secular concerns, either by remaking their surrounding social environment to fit their beliefs or by demarcating boundaries around themselves within which to create a purer group context (Almond et al., 1995). Finally, relative to other forms of change, the above traits together generate a much more overt sense of conflict between fundamentalists and those who do not share their beliefs. As a result, the clashes of knowledge involving fundamentalists are often intense and provocative. Indeed, in an era when religious groups have increasingly supported ecumenicism and interfaith understanding, the confrontational spirit of fundamentalism has become a significant factor in religious conflicts. The following section thus addresses such conflicts in more detail by addressing fundamentalist conceptions of religious certainty and the relationship of those conceptions to fundamentalist clashes with other cultural forces.

Selectivity, Context, and Fundamentalism

As discussed above, fundamentalist movements are distinguished from other religious transformations in part by their greater concern with asserting the authority and authenticity of existing religious certainties than with promoting novel forms of belief. Within this context, however, the distinctiveness of fundamentalist movements also derives from their essential selectivity in defining the fixed core of their system of religious knowledge (Stump, 2000, p. 11). Unlike sects, cults, or schismatic groups, fundamentalists generally do not define their identity in relation to a comprehensive body of religious knowledge. Rather, they emphasize their understanding of the crucial elements of a tradition's worldview and ethos, their notion of the *sine qua non* of true belief. During the Reformation, for example, newly formed branches of Protestant Christianity devised extensive confessions defining the entire range of specific beliefs that they supported. The Westminster Confession of Faith thus contains 33 chapters devoted to Reformed theology, eschatology, ritual, and church organization. Similarly, the Second Helvetic Confession addresses such issues in 30 chapters and the Augsburg Confession in 28. In contrast, Protestant fundamentalists in the United States articulated their core beliefs far more concisely through the so-called Five Points of Fundamentalism:

1. The inerrancy of the Bible
2. The virgin birth of Christ
3. Christ's substitutionary atonement
4. The bodily resurrection of Christ
5. The authenticity of Christ's miracles

The divinity of Christ is sometimes substituted for item (2), and the premillennial Second Coming is sometimes substituted for item (5) (Stump, 2000, p. 28). This limited enumeration of principles did not imply a rejection of other aspects of orthodox Christian belief, of course. Rather, these truths represented the fundamentalists' most vital certainties, and as such they became the focus of clashes of knowledge between fundamentalists and other groups.

An understanding of the role of fundamentalism in clashes of knowledge therefore requires an assessment of the selectivity of fundamentalist articulations of religious knowledge. Most important, this selectivity reflects the defining attitudes of fundamentalists toward the worldview and ethos of their larger tradition. Fundamentalists tend to be highly selective in the importance they assign to different aspects of an orthodox worldview. Protestant fundamentalists, for example, have largely been drawn to beliefs rooted in the literal interpretation of scripture, as suggested in the Five Points listed above. These literalist elements of the worldview, such as the creation or Christ's miracles, have been much more strongly emphasized by fundamentalists than more esoteric questions of theology. Islamic fundamentalism has similarly stressed literal understandings of the Quran, especially with regard to sharia. Fundamentalist movements arising from traditions that lack a concise scriptural tradition may have less literalist inclinations and instead emphasize other core elements of their worldview, such as sacred history among many Hindu fundamentalists or religious identity among Buddhist fundamentalists in Sri Lanka (Stump, 2000, pp. 125–133).

Fundamentalists' worldviews thus represent neither a rejection, nor a radical reinterpretation, nor indeed a precise recreation of established forms of orthodoxy. Their goals are not schismatic, directed toward a broad process of reformation, but they are also not sectarian, espousing a novel understanding of a tradition's religious knowledge. Instead, the objectives of fundamentalists are an expression of cultural resistance, through which they assert their identity as a distinct religious subculture that defines itself in opposition to more conventional religious, social, or cultural groups. As a religious subculture, fundamentalists thus provide an alternate understanding of a tradition's worldview organized around a selected set of orthodox beliefs.

Fundamentalists display a similar selectivity in defining the core elements of their ethos. This selectivity places greatest emphasis on aspects of the ethos that most clearly distinguish a fundamentalist movement from outsiders, creating a sharp boundary between the group and the Other. Fundamentalists thus use their ethos to highlight their distinctive identity, foregrounding group solidarity and opposition to the rest of the world. In linking their identity to group norms of behavior, fundamentalists further use their ethos to assert a dichotomy between their own goodness and the evil of the Other, a distinction that informs subsequent clashes of knowledge.

Fundamentalists reinforce this boundary-making by selectively incorporating an element of the "scandalous" into their ethos: that is, by emphasizing elements that shock or provoke outsiders and that provide a conceptual trap to exclude individuals who are not fully committed to the group (Marty, 1992). In this sense, the fundamentalist ethos also conveys cultural resistance, defining the group as being in conflict with those who do not share its beliefs. Finally, fundamentalists seek to make the distinctiveness of their ethos highly visible outside the group. Such visibility accentuates the boundary delimiting a fundamentalist group, but it also expresses their resistance to other cultural influences and thus tends to provoke overt clashes of knowledge with outsiders. Fundamentalists' interest in broadcasting their distinctiveness is further evinced by their recurring adoption of new technologies to disseminate their views, from radio to cable television to the Internet.

Selectivity thus represents an essential feature of the worldviews and ethoses of fundamentalists and of fundamentalism as a form of cultural resistance. This selectivity also reflects another crucial feature of fundamentalism, its intrinsic contextuality. Fundamentalists' emphasis on certain beliefs and norms within an orthodox worldview and ethos ultimately grows out of their response to conditions within the setting that they inhabit. Fundamentalists in effect selectively constitute their core religious certainties to address concerns linked to their particular geographical context. More specifically, the selectivity of their worldview and ethos derives from a sense of threat, usually understood as a threat to religious truth. Fundamentalism in effect represents a reaction to potential clashes of knowledge, focusing on issues of religious authority and authenticity. Fundamentalists' engagement in such clashes is actually based on their perception of the increasing contextual influence of a perceived threat. Again, this response develops as a contextualized form of resistance aimed at ensuring compatibility between their local setting and their religious certainties.

Although the threats opposed by fundamentalists take diverse forms, they do have some features in common. First, as challenges to religious orthodoxy, such threats are generally unconventional in character. They differ, in other words, from the traditional threats of heresy or heterodoxy, developing instead out of distinctly modern approaches to religious knowledge. Fundamentalists, in turn, believe that these threats require innovative forms of opposition, as expressed in the inherent selectivity of fundamentalism. This belief thus reinforces the modern character of fundamentalism as a form of religious innovation. A second feature of the threats confronted by fundamentalists is that the latter consider them to be pervasive and severe, posing extensive dangers to religious certainty. A related feature is that fundamentalists believe that such threats imperil their own solidarity by potentially undermining the commitment of individual group members. Together, the latter two features account for the vigor and rigidity typical of fundamentalist responses to antagonistic forces. Finally, the threats opposed by fundamentalists are generally local in character, or find expression largely through local manifestations. Fundamentalists thus resort primarily to local action to confront the threats they perceive. The local character of the resulting clashes of knowledge may consequently foster different forms of fundamentalism within the same tradition, each responding to a different contextual threat. This pattern has been particularly evident in contrasts between Islamic fundamentalism in South Asia and the Middle East and also characterizes disparate Buddhist movements in Sri Lanka and Thailand (Stump, 2000, pp. 48–63).

Despite their common features, then, the threats confronted by fundamentalists are rooted in contextual sources within specific settings. The sources of threats to religious certainty thus vary in character, as do the threats that they engender (Stump, 2000, pp. 5–9). Most threats derive from one or more of the four sources outlined below, each a major social trend associated with the circumstances of modernity but realized in specific contexts (see Kong, 2001). These sources do not necessarily represent coherent systems of knowledge in themselves. Instead, they represent broad conceptual frameworks for significant assaults, direct or indirect, on the elements of religious knowledge supported by fundamentalists. These sources of threat, as a result, play key roles in the clashes of knowledge involving fundamentalism and in the grounding of those conflicts in specific settings.

The earliest source of threats for fundamentalists, at least with regard to the American Protestants who first defined a fundamentalist identity, was the rise of modernism within various realms of thought during the 1800s. For Protestant fundamentalists in the United States, modernist reinterpretations of reality and humanity threatened many of the traditional certainties of orthodox belief. They therefore felt obliged to respond to the modern forms of biblical criticism supported by religious liberals and to the novel worldviews and ethoses based on scientific and humanistic concepts. In the process, fundamentalists centered their movement on beliefs that were incompatible with modernism, selectively emphasizing elements of their worldview that focused on the role of supernatural agency. Fundamentalists especially stressed belief in biblical inerrancy, which became the primary focus of their clashes with advocates of modernism. More specifically, belief in the literal truth of the biblical account of creation became the symbolic core of fundamentalist antagonism toward modernism and has remained so into the 21st century.

A second major source of threats for fundamentalists has been the rise of secularism, particularly since the early 20th century. Unlike modernism, secularism does not involve a reinterpretation of religious knowledge. Instead, it downplays the role of religious knowledge as a force in society, especially in public life, relegating faith largely to the private domain. Fundamentalists have seen the rise of secularism as a clear threat to traditional religious certainties and to the previously unchallenged role of religion as a source of authority within society. They have again responded to this threat by selectively defining their primary concerns. In many contexts, fundamentalists have focused on the goal of maintaining society's religious foundations. In these cases, they have usually sought to incorporate selected beliefs into various social structures, most notably legal and educational systems. In the United States, such efforts have focused on restoring prayer to public schools and on restricting the practice of abortion. In India, Hindu fundamentalists have made the passage of national laws limiting the slaughter of cattle a primary objective. Islamic fundamentalists have pursued a broader battle against secularism in some states by advocating the incorporation of sharia into the legal system. Ultra-Orthodox Jewish fundamentalists in Israel have expressed this pattern somewhat differently, focusing on protecting their ability to conform to their ethos within a secular society rather than on the transformation of society at large. Their efforts, too, have reflected a basic selectivity, however. Conflicts over Sabbath-day use of public roads adjacent to ultra-Orthodox neighborhoods have been a central, symbolic concern, for example.

In a smaller number of contexts, pluralism acts as a third major threat to fundamentalist certainties. The threat of pluralism derives from social or governmental acceptance of relativist attitudes toward religion, which in theory treat all religious systems as equals. Pluralistic attitudes do not deny the importance of religion, nor do they recast traditional beliefs in modernist terms, but they still pose a threat to fundamentalists because they can undermine the traditional role of a once hegemonic religion. Pluralism may threaten a once dominant religion, for example, by weakening the religion's social status or its traditional role as a source of authority. Fundamentalists in various contexts have also linked the threat of pluralism to the special role that their religious tradition has played in defining a hegemonic cultural identity. As a result, they have focused on this issue in articulating their central concerns. This pattern has characterized a number of fundamentalist groups in South Asia, for example (Stump, 2000, pp. 63–77). Hindu fundamentalists in India have made Hindutva, or Hinduness, the cornerstone of their movement, emphasizing the role of Hindu culture in shaping Indian society and the importance Hindutva as the basis for the future development of Indian society. In Sri Lanka, Buddhist fundamentalists have similarly organized their concerns around a local concept of Buddhist identity, depicting the traditional Buddhist history of their island as an essential link between the true expression of Theravada Buddhism and their Sinhalese ethnic identity. They have thus continually pushed for greater official status for Buddhism within Sri Lanka as a modern state. In a very different context, an example of the concern with identity and pluralism appeared in the United States in 2007, when the first Muslim elected to Congress chose to be sworn in to his office with his hand placed not on the Bible, as is traditional although not required, but on the Quran. Many Protestant fundamentalists objected to this act, portraying it as a threat to their notion of the Christian identity of the United States.

The influence of colonialism and imperialism in non-Western countries has been a fourth and final source of threats perceived by fundamentalists. It involves the imposition of an alien system of knowledge on a local, traditional society. Through such processes, indigenous varieties of religious knowledge have often acquired secondary status in relation to knowledge systems institutionalized by a colonial power or propagated through informal imperialism. In response, fundamentalists have tended to emphasize indigenous religious certainties that are most at odds with foreign sources of knowledge. Islamic fundamentalists, for example, have often centered their worldview and ethos on either the creation of an authentic Islamic state or on the formal establishment of sharia. The fundamentalist concept of an Islamic state draws on the model of Muhammad's historical empire, of course, but it also represents a modern Islamic response to the lingering effects of colonialism and imperialism in the modern state system. Similarly, fundamentalists' emphasis on sharia reflects orthodox beliefs regarding the law, but it also expresses a reflexive rejection of foreign influences in efforts to create an authentic, indigenous legal system. Efforts by the Muslim Brotherhood during the middle of the 20th century to redefine the Egyptian state in Islamic terms exemplifies the fundamentalist response to this kind of threat, stressing the exclusive legitimacy of indigenous sources of knowledge. The efforts of Islamic fundamentalists over many decades to reconstitute Pakistan in strict Islamic terms reveal a similar emphasis on the authority of Islam in all realms of life.

As a type of religious movement, then, fundamentalism is distinguished by a reflexive and selective emphasis on traditional elements of a worldview and ethos rather than by revolutionary departures from orthodoxy. Fundamentalist movements develop as a modern form of resistance to clashes of knowledge involving threatening influences that originated outside the group's tradition. These influences are largely related to the threats posed by modernism, secularism, pluralism, or colonialism and imperialism, which fundamentalists believe will undermine traditional religious certainties if left unopposed. The emphasis on particular religious certainties expressed by individual fundamentalist movements reflects a selective strategy for countering threats situated within a specific context. The actual effort to eliminate such threats hinges, in turn, primarily on the realization of distinctive responses by fundamentalists, the issue to which the discussion now turns.

Fundamentalism and Clashes of Knowledge

Fundamentalist movements possess an essentially activist character in that they are organized to confront and defeat perceived threats to their systems of religious knowledge. The crucial objective in that process is the articulation, in words and actions, of the religious certainties that most unequivocally challenge a given threat. The resulting confrontation produces a distinct clash of knowledge between the fundamentalist group and those people aligned with the perceived threat. Indeed, such clashes of knowledge lie at the heart of fundamentalism as a contemporary cultural phenomenon. Fundamentalists occasionally contest knowledge with modernist or liberal factions within their own religious tradition. Such clashes tend to focus on questions of authenticity within a common tradition, each side claiming to support the tradition's essential truths. Perhaps more frequently, however, fundamentalists clash with groups and influences originating outside of their religious tradition. Clashes of knowledge in such cases center on the ultimate legitimacy of different forms of knowledge, most commonly setting the fundamentalists' religious certainties in opposition to secular knowledge systems.

Whatever the source of opposition, fundamentalists respond to perceived threats in a clearly reflexive manner. Their contesting of knowledge derives not from a naive attachment to tradition but from a considered engagement with the defense of religious certainty. Their emphasis on specific elements of their worldview and ethos reflect deliberate conceptions of authority and authenticity, but it also informs a conscious delineation of the boundary that separates their group from others. And again, because they focus on a specific threat rather than on more generalized concerns, fundamentalists' involvements in clashes of knowledge center primarily on contextual strategies and goals.

In virtually all cases, fundamentalists seek to counter a threat through the assertion of relevant religious certainties. They undertake that process through diverse strategies, however, depending on the nature of the threat they perceive and on their own underlying beliefs. In some instances fundamentalists draw on their religious system to legitimize their dominance within society, as in the cases of Shiite Islamic fundamentalism in Iran or the Protestant fundamentalists who define the United States as a Christian nation. On the other hand, many fundamentalist groups articulate their religious knowledge to defend their status as a minority within society. They may defend their status by seeking to influence society without immediately dominating it, by openly combating the larger society, or by withdrawing from society into an isolated enclave. The first of these strategies has characterized many Protestant fundamentalists in the United States and less radical Islamic fundamentalists in various Muslim countries. Radical Islamic and Sikh fundamentalists have most obviously followed the second strategy (Stump, 2005). Ultra-Orthodox Jewish fundamentalists in Israel exemplify the third strategy, as does the Santi Asoke sect of Buddhism in Thailand.

In enacting these strategies, fundamentalists pursue goals related to both their worldview and their ethos. Perhaps the primary goal of fundamentalists in clashes of knowledge is to establish the superior truth of their own religious certainties. To realize this goal, they again address the clash of knowledge by emphasizing elements within their worldview that explicitly challenge the threat at the heart of the conflict. Such elements provide the clearest expression of the fundamentalists' understanding of religious truth. Fundamentalists therefore advance these beliefs to assert the supremacy of faith as a way of knowing and to identify the incontestable sources of authority necessary for understanding reality.

Again, the goal of fundamentalists in articulating these religious certainties is neither purely abstract nor universal in scope. Fundamentalists develop their goals in opposition to particular, contextual threats. Their goals thus take on different forms in different settings and clash with competing systems of knowledge in diverse ways. For example, the goal of asserting the authority and inerrancy of the Bible developed among Protestant fundamentalists in response to the rise of modernism in American thought. Within the domain of religion, the latter trend discounted the literal interpretation of the Bible and instead focused biblical criticism on issues of human authorship and textual analysis. In addition, the broader acceptance of modernist perspectives within American society promoted the authority of scientific knowledge, undermining traditional views of authority and truth founded on the Bible. The first major issue through which fundamentalists in the United States clashed with modernist influences, the biblical account of creation, therefore focused directly on the incompatibility between the fundamentalists' worldview and the knowledge generated by modern science or liberal forms of biblical hermeneutics.

Clashes of knowledge involving other fundamentalist groups have been shaped by similar concerns with the supremacy of their worldviews. Islamic fundamentalists in various contexts, for example, have asserted the superiority of their worldview over Western ideas introduced into Muslim regions through the processes of colonialism and imperialism. Islamic fundamentalists have typically focused on the Quranic worldview that defines Islam as a complete way of life requiring submission to Quranic principles in all realms of thought and action. This emphasis has, in turn, led fundamentalists to insist on the authority of the sharia and again to seek the establishment of sharia as the basis for modern law. The particular strategies employed by Islamic fundamentalists have varied, of course. The most radical Egyptian fundamentalists, following the teachings of Sayyid Qutb, have sought to destroy the Egyptian state, which they view as unredeemable, before forging a new state based on the authority of the Quran. In Pakistan, by way of contrast, the major fundamentalist groups have generally sought to remake the existing state through the progressive introduction of Quranic authority into state structures.

Clashes of knowledge involving ultra-Orthodox Jews and the state of Israel similarly have revolved around concepts of authority, with the ultra-Orthodox asserting the primacy of the Torah over secular knowledge. This concern with authority finds its clearest expression in the rejection of the legitimacy of the state of Israel by the most ultra-Orthodox fundamentalists, who assert that the founding of modern Israel was a human attempt to usurp divine authority over the restoration of the Jews to the promised land, part of a process of divine redemption that in their view has not yet begun. A second group of Jewish fundamentalists, the religious Zionists, do not reject the modern state of Israel, but they do oppose its secular character, instead seeing its creation and survival as the beginning of the process of redemption. This group therefore maintains clashes of knowledge with secular Israelis, and with Palestinians, over the meaning of settlement in the West Bank by Jews, including many religious Zionists. For them, possession of territory in the biblical lands of Judea and Samaria represents a crucial expression of the authority of the Torah (Friedman, 1993).

As a final example, Hindu fundamentalists have expressed selective concerns with authority by emphasizing the issue of identity. Their discontent derives from the pluralism expressed in India's constitution. Hindu fundamentalists see this provision as a threat, not to the authority of scripture as in preceding examples but to the authority of tradition and in particular the integrated cultural tradition that has evolved in India over many millennia. Hindu fundamentalists have situated the foundation of that authority in the concept of Hindutva, the sense identity shared by all of those who consider India to be their homeland, their fatherland, and their holy land. While fundamentalists claim that Hindutva expresses nationalist rather than religious meanings, the preceding definition clearly centers on the Indic religions. For Hindu fundamentalists, Hindutva also provides the authoritative basis for a variety of other concerns. This construction of identity as authority thus serves as a central issue in various clashes of knowledge that pit fundamentalists against secularists, moderate Hindus, or non-Hindus.

Fundamentalists, along with their articulation of specific elements of their worldview in relation to sources of authority, selectively advance their ethos as they engage in clashes of knowledge. They do so in part by stressing elements of their ethos that symbolically represent the aspects of their worldview that they simultaneously assert in a clash of knowledge. At the same time, they emphasize the features of their ethos that most explicitly challenge the opposing knowledge system. Again, the fundamentalist emphasis on ethos-based action tends to be highly reflexive and carefully considered, aimed at achieving specific goals in relation to a given, contextual threat. Fundamentalists typically direct such goals toward demonstrating the authenticity of their religious systems or those elements of their ethos rooted in and reflecting religious certainty. They seek to achieve those goals, however, through particular actions organized in response to a local threat, thus reinforcing the spatial connection of the ensuing clash of knowledge to a distinct setting.

In developing specific objectives, fundamentalist movements generally pursue two goals related to the authenticity of their ethos: establishing conformity between existing social structures and the certainties of their faith, and advancing criteria of religious faith and practice that separate true believers from outsiders. Through these goals, fundamentalists seek to assert the legitimacy of their own actions and at the same time establish an identity that stands in sharp contrast to that of persons who do not accept their religious certainties.

Because they embrace a separate identity within society, however, fundamentalists pursue these ethos-related goals from two perspectives, one internal to their group and the other external. Internally, fundamentalists respond to clashes of knowledge by emphasizing their own adherence to their ethos. By conspicuously adhering to their ethos in their daily lives, fundamentalists implicitly assert the authenticity of the ethos itself. In addition, this process reinforces internal conformity to the group's own social structures, conformity that is essential to the group's faith in light of external threats. As an example, Islamic fundamentalist women living in non-Muslim states have often used veiling in recent years as a means of asserting the authenticity of their ethos in contrast to a surrounding dominant culture. The widespread establishment of Christian schools by Protestant fundamentalists in the United States, which allow for school prayer and the teaching of the biblical account of creation, represents the institutionalization of their ethos of faith in biblical truths. Along somewhat different lines, Hindu fundamentalists have stressed the authenticity of their ethos, centering on commitment to Hindu tradition, through their formation of organizations that promote Hindu values. The Rashtriya

Swayamsevak Sangh (National Volunteers Association) ranks as the most important of these, but it has contributed to the formation of many other groups as well (Stump, 2000, pp. 125–131).

In addition to upholding their ethos within their own group, fundamentalists also assert their ethos externally, beyond their group boundaries. They do so in part to further the conformity between external social structures and their own religious certainties. This effort also serves to provoke their opponents, however, and typically becomes overtly confrontational, focusing on the elements of the fundamentalists' ethos that separate them from others. Moreover, fundamentalists implicitly resist any form of compromise with others, and this rigidity in their ethos legitimizes their rejection of other systems of knowledge. The combined effect of these patterns of confrontation, boundary-making, and inflexibility promotes polarization in conflicts between fundamentalists and other groups. In this sense, it is in the provocative extension of the implications of their ethos beyond their own group that fundamentalists most forcefully engage in clashes of knowledge with outsiders.

Again, these clashes of knowledge are highly contextual, shaped both by the perceived dangers confronted by fundamentalists in particular settings and by the selected religious certainties that fundamentalists draw on in addressing those dangers. For ultra-Orthodox Jewish fundamentalists in Israel, for example, an ethos of piety, of observance of Jewish law, and of adherence to traditional views of redemption have expanded to include a confrontational rejection of involvement in the civic life of the state of Israel, a stance that has contributed to an enduring clash with secular society over the meaning of the state and participation in it. In the United States, Protestant fundamentalists have repeatedly adopted a polarizing, activist ethos in seeking to reform American institutions according to their beliefs, as in efforts to expand the role of religion in public schools or to assert the country's Christian identity by introducing various forms of religious expression into government buildings and other forms of public space. In a variety of contexts, and with varying levels of success, Islamic fundamentalists have sought to enforce an inflexible adherence to a strict Islamic ethos by promoting the creation of an authentic Islamic state, based on sharia and governed according to the model of Muhammad's original empire. In pursuing such goals, these fundamentalist groups have devised specific means to redefine the norms and structures of society by asserting the exclusive authority and authenticity of their own system of religious knowledge, and they have consequently provoked ongoing clashes of knowledge with those outside their group.

Conclusions

By definition, fundamentalism is rooted in the concept of religious certainty. Fundamentalist movements arise because their founders believe that they must defend their certainties against some grave form of threat. Because of their unshakable faith in their own worldview and ethos, fundamentalists generally reject the possibility of compromise on matters of religious knowledge. The rejection of compromise indeed represents a key characteristic of fundamentalism in general. As a result, resolutions to the clashes of knowledge involving fundamentalists have proven quite difficult to achieve. Fundamentalists are usually unwilling to cede a degree of legitimacy to the knowledge systems of others, while those opposed by fundamentalists see the latter as hopelessly strict and unyielding in their way of thinking.

Nonetheless, clashes of knowledge involving fundamentalists may ultimately achieve some type of resolution through the effects of continuing processes of culture change. Fundamentalism is not a fixed form of religious expression. It instead represents a particular cultural phenomenon that has arisen as traditional faiths have encountered specific forms of social change in the modern era. As a form of cultural expression, fundamentalism will thus continue to evolve in relation to the particular contexts where it has developed. Like other elements of culture, moreover, fundamentalist movements will not be perfectly reproduced over time by their adherents. They will repeatedly incorporate new patterns of thought and behavior introduced through the changing perspectives of their adherents and the changing influence of their surroundings. One possible outcome of this process may be gradual forms of accommodation to the rest of society. This trend appears to some extent in Protestant fundamentalist efforts to establish an alternative to the teaching of evolution, which has developed from a simple emphasis on the biblical account of creation to the more complex explanations of creationism and intelligent design, which unite biblical teachings with purportedly scientific methods. Trends toward accommodation may eventually lead fundamentalist groups to accept ideas external to their own traditions. As an example, in recent years as Islamic fundamentalists in Turkey have responded to government bans on women wearing headscarves in government buildings and universities, they have done so not in religious terms but rather in secular terms by casting this controversy as an issue of human rights (Stump, 2004). This approach, defending an Islamic practice through reference to the modernist ethos of human rights instead of through a provocative insistence on a strict Islamic ethos, signifies a notable development in how fundamentalists have sought to resolve clashes of knowledge.

A second potential outgrowth from present fundamentalist patterns involves the development of less confrontational forms of reflexive religious traditionalism. Such movements build on the success that earlier fundamentalist groups had in establishing the modern vitality of religious traditionalism, contrary to the once widely assumed ascendancy of secularization (Swatos & Cristiano, 1999). These "postfundamentalist" movements differ from earlier expressions of fundamentalism in significant ways, however. They tend to be more sectarian and thus emphasize narrower group identities. They also replace the somewhat simpler, generic enunciation of fundamental principles, typical of earlier movements, with a more extensive and detailed articulation of the whole of a religious system. Postfundamentalist movements tend as well to be less exclusively tied to concerns within a particular context, focusing at least in part on the wider implications of their worldview and ethos. As a result, postfundamentalists generally display less specific concerns with immediate, contextual threats. Instead, they focus on the inevitable unfolding of sacred history, adopting a teleological outlook that lies at the heart of their distinctive identity and system of knowledge. This perspective minimizes the significance of local threats, viewing them as being irrelevant to the larger course of sacred history. A prominent example of postfundamentalism has emerged in the Christian Reconstructionist movement in the United States, dating from the 1960s, which combines a strict Calvinist ethos with a postmillennial view of sacred history. Although fundamentalist in its origins, this group pays less attention to local threats than to the completion of sacred history in the Second Coming. As a result, although this group's knowledge system differs substantially from that of mainstream U.S. society, it has engaged in fewer clashes of knowledge with outsiders than earlier fundamentalist groups have. The rise of this group and similar ones in other traditions may suggest a possible lessening of conflicts involving religious knowledge even as some religious groups remain distinctly traditional in their values and beliefs.

In sum, the contesting of knowledge by fundamentalist groups is itself a contextual cultural phenomenon, the product of the historical and geographical intersection of increasingly reflexive religious traditionalists and a variety of threatening social changes. Fundamentalists' encounters with the rise of modernism, of secularism, of pluralism, and of the cultural implications of colonialism and imperialism have all contributed to the formation of diverse milieus of discord in which new social perspectives and structures have clashed with traditional religious certainties. At the same time, such patterns of discord are likely to undergo the same processes of change that characterize all expressions of culture. The ongoing reproduction of contemporary societies will thus transform, at least to some extent, the clashes of knowledge involving fundamentalists. So, too, will changes within fundamentalist groups themselves, through their gradual accommodation of certain mainstream beliefs or their division into more exclusive postfundamentalist movements less concerned with specific threats.

References

Almond, G. A., Sivan, E., & Appleby, R. S. (1995). Explaining fundamentalism. In M. Marty & R. S. Appleby (Eds.), *Fundamentalisms comprehended* (pp. 425–444). Chicago, IL: University of Chicago Press.

Badone, E. (Ed.). (1990). *Religious orthodoxy and popular faith in European society*. Princeton, NJ: Princeton University Press.

Charlesworth, M. (1997). *Religious inventions: Four essays*. Cambridge, England: Cambridge University Press.

Conkin, P. K. (1997). *American originals: Homemade varieties of Christianity*. Chapel Hill, NC: University of North Carolina Press.

Friedman, M. (1993). Jewish zealots: Conservative versus innovative. In L. J. Silberstein (Ed.), *Jewish fundamentalism in comparative perspective: Religion, ideology, and the crisis of modernity* (pp. 148–163). New York: New York University Press.

Geertz, C. (1973). *The interpretation of cultures*. New York: Basic Books.

Kong, L. (2001). Mapping "new" geographies of religion: Politics and poetics in modernity. *Progress in Human Geography, 25*, 211–233.

Marty, M. E. (1992). Fundamentals of fundamentalism. In L. Kaplan (Ed.), *Fundamentalism in comparative perspective* (pp. 15–23). Amherst, MA: University of Massachusetts Press.

Marty, M. E., & Appleby, R. S. (Eds.). (1991–1995). *The fundamentalism project* (Vols. 1–5). Chicago, IL: University of Chicago Press.

Nolan, M. L., & Nolan, S. (1992). *Christian pilgrimage in modern Western Europe*. Chapel Hill, NC: University of North Carolina Press.

Pals, D. L. (1996). *Seven theories of religion*. New York: Oxford University Press.

Smith, J. Z. (1996). A matter of class: Taxonomies of religion. *Harvard Theological Review, 89*, 387–403.

Stark, R., & Bainbridge, W. S. (1987). *A theory of religion*. New York: Peter Lang.

Stump, R. W. (2000). *Boundaries of faith: Geographical perspectives on religious fundamentalism*. Lanham, MD: Rowman & Littlefield.

Stump, R. W. (2004). Fundamentalism, democracy, and the contesting of meaning. In D. Odell-Scott (Ed.), *Democracy and religion: Free exercise and diverse visions* (pp. 185–201). Kent, OH: Kent State University Press.

Stump, R. W. (2005). Religion and the geographies of war. In C. Flint (Ed.), *The geography of war and peace* (pp. 149–173). Oxford, England: Oxford University Press.

Swatos, W. H., Jr., & Cristiano, K. J. (1999). Secularization theory: The course of a concept. *Sociology of Religion, 60*, 209–228.

Chapter 11
The Theory of Cognitive Dissonance: State of the Science and Directions for Future Research

Peter Fischer, Dieter Frey, Claudia Peus, and Andreas Kastenmüller

The theory of cognitive dissonance is one of the most influential theories in social psychology. Since its initial publication 50 years ago, it has inspired more than 1,000 empirical papers. However, dissonance theory has not only had a profound impact on research in social psychology, it has also been used for designing interventions to address a variety of societal problems. In this short overview of the empirical literature on dissonance theory, we first introduce the definition of dissonance theory in its classic formulation by Festinger (1957). Second, we review the most important paradigms used in empirical dissonance research and summarize the most prominent empirical results. Third, we present the main features of the self-based revision of dissonance theory and introduce our own self-based modification of dissonance theory including related data on ego-depletion and selective exposure. Finally, we present directions for future dissonance research, in particular in the areas of self-regulation and information-processing, and discuss the application of dissonance theory to societal problems.

Classic Formulation of Dissonance Theory

Cognitive dissonance is defined as the subjective perception of incompatibility between two self-relevant cognitions. A cognition can be any element of knowledge, belief, attitude, value, emotion, interest, plan, or behavior. In other words, cognitions are dissonant when one specific cognition implies the opposite of another cognition. The resulting cognitive discrepancy is associated with a psychological state of unpleasantness (cognitive dissonance) that motivates the individual to reduce this state of discomfort by reducing the discrepancy between the dissonant cognitions (Festinger, 1957; Harmon-Jones, 2000). The magnitude of the cognitive dissonance is determined by the importance of the cognitions involved and their relation to a personal standard. Dissonance can be reduced in five ways or some combination thereof: (a) adding consonant cognitions, (b) subtracting dissonant cognitions (by ignoring, suppressing, or forgetting them), (c) replacing existing cognitions with others, that is, subtracting dissonant cognitions while adding consonant ones, (d) increasing the importance of consonant cognitions, and

P. Meusburger et al. (eds.), *Clashes of Knowledge*.

(e) reducing the importance of dissonant cognitions. Adding consonant cognitions can also be described as a justification process, and reducing the importance of inconsistent information is often found in trivialization processes.

Classic dissonance research has largely been based on three types of paradigms: (a) induced compliance, (b) free choice, and (c) selective exposure. The induced-compliance paradigm (e.g., Festinger & Carlsmith, 1959) involves asking participants to engage in behavior that is counter to their personal opinion or preference (e.g., performing a dull writing task). Afterwards, participants are urged to lie to a fellow participant by describing the task as very interesting. In this classical experiment the dissonance-inducing lying behavior was either performed in exchange for a low reward of 1 dollar (low justification) or a high reward of 20 dollars (high justification). The dependent variable that Festinger and Carlsmith measured was the participant's attitude toward the dull task that he or she had worked on. The classic result was that participants with low justification for lying rated the dull task as more interesting than participants with high justification did. The two researchers explained this finding with a dissonance-reduction process, contending that participants in the 1-dollar condition were less able to attribute their lying (dissonant behavior) to the financial reward they received than were the participants in the 20-dollar condition. Overall, studies based on the induced-compliance paradigm have shown that people who have exhibited a certain behavior that contrasts their actual opinion reduce the resulting dissonance by changing their attitude. This effect is less pronounced when the behavior can be justified otherwise, as by high rewards.

The free-choice paradigm typically manipulates dissonance arousal by means of different levels of decision difficulty (e.g., Brehm, 1956). For example, participants are asked to rank different consumer goods and afterwards are instructed to decide between the consumer good ranked second and the one ranked sixth (low dissonance) or between the one ranked second and that ranked third (high dissonance). Subsequently, the participants are asked to indicate the desirability of the two goods. The classic finding for the high-dissonance condition is the spreading-apart-of-alternatives effect, describing the fact that the chosen good increases in desirability whereas the nonchosen good decreases in desirability.

The third prominent paradigm employed in research on cognitive dissonance is the selective exposure to information (see Frey, 1986; Jonas et al., 2001). Typically, dissonance is induced by a difficult decision participants have to make (e.g., between two equally attractive consumer goods, investment strategies, or political plans). Afterwards, they receive additional information (normally between 8 and 16 pieces) of which half support and half contradict the participant's previous decision. Participants are then asked to select those pieces of information they want to read about in greater detail. Within this dissonance paradigm, the classic finding is a confirmation bias, that is, participants normally select significantly more decision-consistent than decision-inconsistent pieces of information. The information-search paradigm is of particular practical relevance because several studies have provided evidence that neglecting decision-inconsistent and focusing on decision-consistent information is associated with poor decision outcomes (e.g., Janis, 1982; Kray & Galinski, 2003; Schulz-Hardt, 1997).

Several empirical studies based on the classic formulation of dissonance theory have investigated which conditions affect the degree to which individuals recognizably engage in dissonance reduction. In summary, their results show that individuals with high levels of commitment to a certain behavior or standpoint exhibit more pronounced dissonance-reduction effects than do individuals with lower corresponding commitment. For instance, Brock and Balloun (1967) found that smokers were more liable than nonsmokers to neglect health information that is inconsistent with smoking (see also Feather, 1962). Other studies revealed that high dissonance is elicited only under conditions of high subjective choice. For example, Frey and Wicklund (1978) demonstrated that confirmation bias in information search is stronger when participants had made the decision under high- rather than low-choice conditions.

An early study by Nel et al. (1969) revealed another factor that has an impact on the degree to which people reduce dissonance. The authors reported that they had observed a dissonance effect (attitude change) only when participants had expected their attitude-inconsistent behavior, in this case publicly proposing to legalize marihuana, to affect other people negatively (see also Cooper & Fazio, 1984). Rhine (1967) found a curvilinear relationship between the level of dissonance arousal and dissonance reduction that follows an inverse U-function. Specifically, individuals tend to increasingly reduce dissonance until reaching a maximum point; when dissonance arousal reaches a critical level, individuals decrease their dissonance-reduction efforts in order to prepare a change in attitude, decision, or standpoint. The empirical results presented above help one understand a variety of non-common-sense phenomena that can be explained by dissonance theory. For example, the predictions of the theory can explain why dissonance reduction is stronger under conditions of low punishment than of high punishment (forbidden toy paradigm, Aronson & Carlsmith, 1963), why attitude change is stronger under conditions of low reward than of high reward (the $1/$20 experiment by Festinger and Carlsmith, 1959), and why the attractiveness of a decision alternative or standpoint increases with the extent to which a person has previously invested in this decision or standpoint (escalation of commitment, Aronson, 1961).

More recent findings in dissonance research address the moderating role of personality on the motivation to reduce dissonance, dissonance and the integration of knowledge, and the application of dissonance theory to societal phenomena. With regard to personality attributes, it has been shown that people with a high need for closure (Kruglanski, 1989) have a greater tendency to reduce dissonance than do people with a low need for closure (see also Fischer et al., 2007a). Furthermore, individuals with high cognitive complexity are less motivated to reduce dissonance than those with low cognitive complexity (Harvey, 1965). Finally, individuals with high attributional complexity (i.e., high ability to find external justifications of their own behavior) show less dissonance reduction (attitude change) than people low in attributional complexity (Stalder & Baron, 1998).

With regard to integration of knowledge, for instance, Festinger et al. (1956) investigated a doomsday cult whose members were convinced the earth was going to blow up. However, when the predicted date of the cataclysm had passed and the earth had not ceased to exist, the members of the cult bolstered their belief system (active change of knowledge), attributed the planet's survival to the power of their prayers, and tried to find new cult members. In another line of research, Janis (1982) found that members of an advisory board around President Kennedy in 1961 referring to an imminent attack on Cuba neglected information that was inconsistent to the opinion of the whole group of advisors. The author found that group members experience dissonance when they realize that other group members have different opinions; subsequently, they try to reduce dissonance by persuading other group members of their opinion, urging consensus, or changing their own position (Matz & Wood, 2005). The integration of knowledge is particularly important in politics. In the context of the Watergate affair, for example, Sweeney and Gruber (1984) found that conservative voters in the United States were more inclined to neglect information that was inconsistent with their political position than liberal voters were (for a similar effect, see Jonas et al., 2003).

Finally, dissonance processes are also relevant for the understanding of interpersonal and societal processes. For example, researchers found cultural differences in the way collectivistic (Asian Canadians) and individualistic (European Canadians) individuals justified their choices. More specifically, collectivists justified their choices more when they had made a decision for a friend than when they had made a decision for themselves, whereas individualists justified their decision more strongly when they had made it for themselves than when they had made it for a friend (see Hoshino-Browne et al., 2005). In addition, studies have revealed a phenomenon called "vicarious dissonance," the subjective perception of incompatibility experienced by individuals who have witnessed members of important in-groups engage in inconsistent behavior. Vicarious dissonance also leads the perceivers to experience dissonance and thus change their attitudes. The mediating mechanism has been found to be the discomfort that observers imagined they would feel if they were in the actor's place (see Norton et al., 2003). In another interpersonal context, McGregor et al. (2001) found that personal uncertainty (caused by a threat to self-integrity) arouses dissonance and, in turn, promotes authoritarianism, a hardening of attitude, and the devaluation of out-groups.

Modifications, New Formulations, and Self-Based Revisions of Dissonance Theory

As for many other theories, it has been questioned whether dissonance theory is a more motivational or cognitive theory. The motivational formulation of dissonance theory is supported by the finding that dissonance indeed is associated with physiological arousal (Elkin & Leippe, 1986). It also has properties of general arousal, meaning that high dissonance increases performance on simple tasks but reduces performance on difficult tasks (for a review, see Kiesler & Pallak, 1976). In addition, dissonance is experienced as psychological discomfort, as documented by Elliot & Devine (1994), who showed that dissonance is a distinct aversive feeling instead of an undifferentiated general arousal state. (They also provided a self-report questionnaire for measuring dissonance arousal.) Along this line of argumentation, Cooper and Fazio (1984) formulated their "new look model" by distinguishing between dissonance motivation and dissonance arousal. They stated that dissonance arousal is characterized as an undifferentiated physiological arousal (which can be labeled positively or negatively). Dissonance motivation results and the typical dissonance effects can be observed in individuals only if this arousal state is labeled negatively.

Another very influential revision of dissonance theory addresses the relation between dissonance arousal and the involved self. According to this revision, dissonance is aroused only when people act in ways that are inconsistent with their core beliefs and thus their self (Aronson, 1968, 1999). Accordingly, the author also derived from this assumption that dissonance arises not because of mere cognitive inconsistency but because of cognitions causing self-inconsistency. This self-based revision of dissonance theory holds that dissonance is aroused in the experiment by Festinger and Carlsmith (1959) because of the discrepancy between "I am an honest person" and "I lie to fellow students" and not so much because the cognitions "I said the task was exciting" and "Indeed, the task was boring" are incompatible. In summary, authors like Aronson (1999) and Harmon-Jones (2000) argue that dissonance is aroused because of threats to a person's positive self-conceptions. Several studies support the validity of the first self-based revision of dissonance theory. For example, Stone et al. (1994) conducted hypocrisy experiments in which participants gave a persuasive speech advocating safe sex. This speech was given either publicly (in front of a video camera) or privately (without being videotaped). The second experimental factor was whether a past failure to use condoms was made salient or not. The dependent variable the authors measured was the intention to practice safe sex in the future (participants could purchase condoms with their experimental reward). The main result of the study was that individuals in the hypocrisy condition (public speech *and* high salience of past failure to use condoms) purchased more condoms than participants of all remaining three conditions.

Finally, a second self-based revision of dissonance theory was set forth by the self-affirmation theorists. According to this theoretical perspective, dissonance effects do not occur because of cognitive inconsistency but because of the need or motivation to maintain an overall image of self-integrity (e.g., Steele & Liu, 1983; Steele et al., 1993; see also Harmon-Jones, 2000). Hence, in typical dissonance situations individuals do not change their attitude because of cognitive discrepancy or self-inconsistency but because of their need or motivation to maintain a positive self-image. Freely behaving in contradiction to one's core attitudes or making difficult decisions threatens the positive self-image, whereas the affirmation of important aspects of the self-concept helps maintain or restore self-integrity. Empirical support for the validity of this self-affirmation perspective on dissonance processes was provided by Steele (1988), who did not find the typical dissonance-related attitude change when participants had been given the possibility to affirm their global self-integrity (by expressing an important self-relevant value in an essay) prior to their behavior. In addition, Tesser and Cornell (1991) found that the increased salience of positive self-evaluations also decreases the motivation to reduce dissonance.

Both self-based revisions make valid predictions but contradict themselves in specific aspects. For example, the self-consistency revision predicts increased dissonance reduction for individuals with high self-esteem (i.e., high self-esteem should increase the discrepancy between attitude and attitude-inconsistent behavior). In contrast, the self-integrity revision proposes the opposite: decreased dissonance-reduction effects for individuals with high self-esteem (i.e., high self-esteem buffers against threatening dissonance arousal). Moreover, the previous two self-based revisions of dissonance theory contain a rather passive role of the self, which either (a) functions as a reference point (self-consistency revision) for comparing one's own counter-attitudinal behavior with one's core values and attitudes or (b) represents a cognitive meta-structure that motivates the individual to maintain self-integrity through self-affirmation. Neither revision makes a statement about the active agent in this process. We propose that the self has a more dynamic role in the dissonance-reduction process than has been assumed in the self-consistency and self-affirmation theory. We present a short outline of our theoretical argument and first empirical findings on ego-depletion and dissonance-reduction processes in the following paragraph.

Self-Regulation and Dissonance: The Impact of Ego-Depletion on Confirmatory Information-Processing

A theoretical perspective that has been developed in recent years further supports the self-integrity (self-affirmation) revision of dissonance theory. Within this perspective, self-regulation is regarded as a process of a person's conscious will. For example, self-regulation is required when a person tries to abstain from eating while dieting. In general, self-regulation is required when a person tries to override spontaneous cognitive, affective, or behavioral responses (see Muraven & Baumeister, 2000; Schmeichel et al., 2003). This process of self-control is defined as the exertion of control over the self by the self (Baumeister, 1998). Hence, the self has only limited self-regulatory strength, which can be regarded as some form of power or energy. If a person uses her or his self-regulatory resources (e.g., by controlling thoughts, emotions, or behaviors), the amount of this energy is reduced (until the energy is replenished). Several lines of recent research have revealed that self-regulatory resources are involved in a variety of processes and behaviors, including higher intellectual performance, interpersonal processes (impression management), inhibition of aggression, or decision-making and information-processing (e.g., Baumeister, 1998; Fischer et al., 2007b; Muraven & Baumeister, 2000; Schmeichel et al., 2003). Typically, participants perform a self-regulation task (e.g., controlling attention) and are subsequently asked to perform another self-regulation task. The typical result on the second regulatory task is that ego-depleted participants (those who had performed a self-regulation task) are outperformed by nondepleted participants.

Applying this theoretical perspective to dissonance processes, we propose that self-regulatory resources are also required when individuals have to abstain from dissonance-reduction processes. In other words, we predict that ego-depleted participants should have less self-affirming resources for abstaining from dissonance-reduction processes than nondepleted participants do. We tested this proposition by using a classic information-search paradigm (selective exposure). Four studies in two of our manuscripts (Fischer et al., 2007a, b) employing political and economic decision-making scenarios consistently demonstrated that individuals with depleted regulatory resources exhibit a stronger tendency for confirmatory information-processing than nondepleted individuals do. Mediational analyses suggested that individuals with depleted regulatory resources cling to their standpoint more strongly and find inconsistent information to be more unpleasant and aversive than is the case with their nondepleted counterparts and that this dissonance leads to increased confirmatory information-processing. Ego threat, cognitive load, and other explanations for the effect of ego-depletion on confirmatory information-processing were thus ruled out. In summary, this set of studies constitutes initial evidence for the assumption that self-regulatory resources are required in order to resist dissonance-reduction tendencies, such as selective exposure and confirmatory information-processing. Therefore, the self might be a more active agent in dissonance processes than has been assumed in previous self-based revisions of dissonance theory.

Conclusion and Future Perspectives

Cognitive discrepancies are associated with dissonance—an aversive motivational state that occurs mainly when individuals behave counter-attitudinally or make difficult decisions. The main routes of dissonance reduction are (a) attitude change, (b) trivialization, and (c) search for supporting information. These processes are used to justify prior behavior, so it can be concluded that humans are not rational but rationalizing. The self plays an extraordinarily important role in understanding dissonance effects. Significant revisions of dissonance theory are set forth by self-based theories, that is, by self-consistency theory and self-affirmation theory. We have also learned that dissonance theory is a universal theory but that the specific culture determines what is dissonant and what is consonant. Dissonance theory pertains to the individual level, but it also makes valid predictions at a group level. In short, dissonance theory is a powerful social psychological theory that can be employed to explain many social phenomena, such as extremism or barriers to societal change.

However, even after 50 years of dissonance research and more than 1,000 publications, there are still many open questions about the impact of cognitive discrepancy on human cognition, emotion, and behavior. A fruitful endeavor for future research might be to resolve the conflicting predictions between self-consistency and self-affirmation theory. In addition, researchers should also try to better clarify the dynamic role of the self in dissonance-reduction processes. Our own studies (Fischer et al., 2007a, b) are a starting point in this direction.

From a practical perspective, dissonance theory is a powerful theoretical tool with which to understand and predict striking social issues. It can explain why people who are committed to a certain value, ideology, or theory are relatively closed-minded and why they selectively seek information that supports their views. As a consequence, existing stereotypes are sustained. Dissonance theory thus provides an explanation for the fact that people often find it difficult to tolerate the norms and values of other people, which frequently results in conflict. The question is how this closed-mindedness can be overcome. One way may be to demonstrate that closed-mindedness and selective search for information is dissonant to even higher-order values, such as openness to new information or cosmopolitan values, and to global goals. Dissonance theory can also add to an understanding of why people are so reluctant to tackle many of the severe problems the world will face in the coming years, such as global warming, the shortage of water, and overpopulation. These problems are very threatening and arouse high levels of dissonance, which leads to the selective search for information that euphemizes the problems. As Festinger (1957) emphasized, however, there is a short run and a long run to dissonance reduction. Strategies that reduce dissonance in the short run may not do so in the long run. He stressed that, in order to develop future perspectives, people have to enlarge their narrow views and explicitly search for dissonant information. This mindset derived from empirical findings on dissonance theory is the cornerstone of a tolerant society that actively addresses the problems it is faced with.

References

Aronson, E. (1961). The effect of effort on the attractiveness of rewarded and unrewarded stimuli. *Journal of Abnormal and Social Psychology, 63*, 375–380.

Aronson, E. (1968). Dissonance theory: Progress and problems. In R. P. Abelson, E. Aronson, W. J. McGuire, T. M. Newcomb, M. J. Rosenberg, & P. H. Tannenbaum (Eds.), *Theories of cognitive consistency: A sourcebook* (pp. 5–27). Chicago, IL: Rand McNally.

Aronson, E. (1999). Dissonance, hypocrisy, and the self-concept. In E. Harmon-Jones & J. Mills (Eds.), *Cognitive dissonance: Progress on a pivotal theory in social psychology* (pp. 103–126). Washington, DC: American Psychological Association.

Aronson, E., & Carlsmith, J. M. (1963). Effect of the severity of threat on the devaluation of forbidden behaviour. *Journal of Abnormal and Social Psychology, 66*, 584–588.

Baumeister, R. F. (1998). The self. In D. Gilbert, S. T. Fiske, & G. Lindzey (Eds.), *Handbook of social psychology* (4th ed., pp. 680–740). Boston, MA: McGraw-Hill.

Brehm, J. W. (1956). Postdecision changes in the desirability of alternatives. *Journal of Abnormal and Social Psychology, 52*, 384–389.

Brock, T. C., & Balloun, J. L. (1967). Behavioral receptivity to dissonant information. *Journal of Personality and Social Psychology, 6*, 413–428.

Cooper, J., & Fazio, R. H. (1984). A new look at dissonance theory. In L. Berkowitz (Ed.), *Advances in experimental social psychology* (Vol. 17, pp. 229–262). Hillsdale, NJ: Erlbaum.

Elkin, R. A., & Leippe, M. R. (1986). Physiological arousal, dissonance, and attitude change: Evidence for a dissonance-arousal link and a "don't remind me" effect. *Journal of Personality and Social Psychology, 51*, 55–65.

Elliot, A. J., & Devine, P. G. (1994). On the motivational nature of cognitive dissonance: Dissonance as psychological discomfort. *Journal of Personality and Social Psychology, 67*, 382–394.

Feather N. T. (1962). Cigarette smoking and lung cancer: A study of cognitive dissonance. *Australian Journal of Psychology, 14*, 55–64.

Festinger, L. (1957). *A theory of cognitive dissonance*. Evanston, IL: Row, Peterson.

Festinger, L., & Carlsmith, J. M. (1959). Cognitive consequences of forced compliance. *Journal of Abnormal and Social Psychology, 58*, 203–210.

Festinger, L., Riecken, H., & Schachter, S. (1956). *When prophecy fails*. Minneapolis, MN: University of Minnesota Press.

Fischer, P., Frey, D., & Jonas, E. (2007a). Evidence that need for closure increases selective exposure to supporting information. Manuscript submitted for publication.

Fischer, P., Greitemeyer, T., & Frey, D. (2007b). Ego-depletion and selective exposure: The impact of self-regulatory resources on confirmatory information processing. Manuscript submitted for publication.

Frey, D. (1986). Recent research on selective exposure to information. In L. Berkowitz (Ed.), *Advances in experimental social psychology* (Vol. 19, pp. 41–80). San Diego, CA: Academic.

Frey, D., & Wicklund, R. (1978). A clarification of selective exposure: The impact of choice. *Journal of Experimental Social Psychology, 14*, 132–139.

Harmon-Jones, E. (2000). An update on dissonance theory, with a focus on the self. In A. Tesser, R. Felson, & J. Suls (Eds.), *Psychological perspectives on self and identity* (pp. 119–144). Washington, DC: American Psychological Association.

Harvey, O. J. (1965). Some situational and cognitive determinants of dissonance resolution. *Journal of Personality and Social Psychology, 1*, 349–355.

Hoshino-Browne, E., Zanna, A. S., Spencer, S. J., Zanna, M. P., Kitayama, S., & Lackenbauer, S. (2005). On the cultural guises of cognitive dissonance: The case of easterners and westerners. *Journal of Personality and Social Psychology, 89*, 294–310.

Janis, I. L. (1982). *Groupthink* (2nd rev. ed.). Boston, MA: Houghton Mifflin.

Jonas, E., Schulz-Hardt, S., Frey, D., & Thelen, N. (2001). Confirmation bias in sequential information search after preliminary decisions: An expansion of dissonance theoretical research on "selective exposure to information." *Journal of Personality and Social Psychology, 80*, 557–571.

Jonas, E., Graupmann, V., Fischer, P., Greitemeyer, T., & Frey, D. (2003). Dissonanz als Wahlkampfhelfer—Konfirmatorische Informationssuche im Kontext der Parteispendenaffäre der CDU [Dissonance as a supporter in an election campaign—Selective exposure in the context of the party donation affair of the CDU]. *Zeitschrift für Sozialpsychologie, 34*, 47–61.

Kiesler, C. A., & Pallak, M. S. (1976). Arousal properties of dissonance manipulations. *Psychological Bulletin, 83*, 1014–1025.

Kray, L. J., & Galinsky, A. D. (2003). The debiasing effect of counterfactual mindsets: Increasing the search for disconfirmatory information in group decisions. *Organizational Behavior and Human Decision Processes, 91*, 69–81.

Kruglanski, A. W. (1989). *Lay epistemic and human knowledge: Cognitive and motivational bases*. New York: Plenum.

Matz, D., & Wood, W. (2005). Cognitive dissonance in groups: The consequences of disagreement. *Journal of Personality and Social Psychology, 88*, 22–37.

McGregor, I., Zanna, M. P., Holmes, J. G., & Spencer, S. J. (2001). Compensatory conviction in the face of personal uncertainty: Going to extremes and being oneself. *Journal of Personality and Social Psychology, 80*, 472–488.

Muraven, M., & Baumeister, R. F. (2000). Self-regulation and depletion of limited resources: Does self-control resemble a muscle? *Psychological Bulletin, 126*, 247–259.

Nel, E., Helmreich, R., & Aronson, E. (1969). Opinion change in the advocate as a function of the persuasibility of his audience: A clarification of the meaning of dissonance. *Journal of Personality and Social Psychology, 12*, 117–124.

Norton, M. I., Monin, B., Cooper, J., & Hogg, M. A. (2003). Vicarious dissonance: Attitude change from the inconsistency of others. *Journal of Personality and Social Psychology, 85*, 47–62.

Rhine, R. J. (1967). Some problems in dissonance theory research on information selectivity. *Psychological Bulletin, 68*, 21–28.

Schmeichel, B. J., Vohs, K. D., & Baumeister, R. F. (2003). Intellectual performance and ego depletion: Role of the self in logical reasoning and other information processing. *Journal of Personality and Social Psychology, 85*, 33–46.

Schulz-Hardt, S. (1997). *Realitätsflucht in Entscheidungsprozessen: Vom Groupthink zum Entscheidungsautismus* [Escape from reality in decision-making processes: From groupthink to the autism of decisions]. Bern: Huber.

Stalder, D., & Baron, R. S. (1998). Attributional complexity as a moderator of dissonance-produced attitude change. *Journal of Personality and Social Psychology, 75*, 449–455.

Steele, C. M. (1988). The psychology of self-affirmation: Sustaining the integrity of the self. In L. Berkowitz (Ed.), *Advances in experimental social psychology* (Vol. 21, pp. 261–302). San Diego, CA: Academic.

Steele, C. M. & Liu, T. J. (1983). Dissonance processes as self-affirmation. *Journal of Personality and Social Psychology, 45*, 5–19.

Steele, C. M., Spencer, S. J., & Lynch, M. (1993). Self-image resilience and dissonance: The role of affirmational resources. *Journal of Personality and Social Psychology, 64*, 885–896.

Stone, J., Aronson, E., Crain, A. L., Winslow, M. P., & Fried, C. B. (1994). Inducing hypocrisy as a means for encouraging young adults to use condoms. *Personality and Social Psychology Bulletin, 20*, 116–128.

Sweeney, P. D., & Gruber, K. L. (1984). Selective exposure: Voter information preferences and the Watergate affair. *Journal of Personality and Social Psychology, 46*, 1208–1221.

Tesser, A., & Cornell, D. P. (1991). On the confluence of self processes. *Journal of Experimental Social Psychology, 27*, 501–526.

Chapter 12
Turning Persuasion from an Art into a Science

Robert B. Cialdini

What is the place of the persuasion process in the topic of clashes of knowledge? The outcome of such clashes is often determined not so much by the features of the knowledge itself as by the features of the way the knowledge is presented. Having a good case to make is not enough. It is the side that makes its good case well (i.e., most persuasively) that will frequently win the day. The focus of this chapter, then, will be on methods for communicating one's case in the most effective manner so as to prevail in clashes of knowledge.

The Roots of Persuasion Studies

Dangerous Fruit

First, a brief step into the past is in order. The renowned scholar of social influence, William McGuire, determined that in the four millennia of recorded Western history, there have been only four scattered centuries in which the study of persuasion flourished as a craft. The first was the Periclean Age of ancient Athens; the second occurred during the years of the Roman Republic; the next appeared in the time of the European Renaissance; and the last extended over the 100 years that have just ended and that witnessed the advent of large-scale advertising, information, and mass-media campaigns (McGuire, 1985). Although this bit of background seems benign, it possesses an alarming side: Each of the three previous centuries in the systematic study of persuasion ended similarly when political authorities had the masters of persuasion killed.

A moment's reflection suggests why this pattern occurred. Information about the persuasion process was dangerous because it created a base of power entirely separate from that which the authorities of the times controlled. Persuasion is a way to move people that does not require coercion, intimidation, or brute strength. Persuaders win the day by marshalling forces that heads of state have no monopoly over, such as cleverly crafted language, properly placed information, and, most important, psychological insight. To eliminate this rival source of influence, it was easiest for

P. Meusburger et al. (eds.), *Clashes of Knowledge*.

the rulers to eliminate those few individuals who truly understood how to engage in the process.

Consequently, each of the three earlier centuries in the systematic study of persuasion ended in the same unsettling manner—with a purge of the reigning persuasion experts. It has not been long since the completion of the fourth such century. Therefore, should those who study and master the material contained in this chapter begin looking for cover out of fear that they might be included in an impending fourth day of annihilation? Not this time.

The Flowering of Science

Something revolutionary has happened to the study of persuasion during the past half century. In the bargain, the change has rendered implausible the idea that persuasion expertise could be eradicated by eradicating the persuasion experts. Alongside the art of persuasion has grown a formidable science of the process. For over 50 years, researchers have been applying a rigorous scientific approach to the question of which messages most successfully lead people to concede, comply, or change. Under controlled conditions, they have documented the sometimes astonishing impact of making a request in one fashion versus making the identical request in a slightly different fashion. Besides the sheer size of the effects these researchers have uncovered, there is another noteworthy aspect of their results—they are repeatable.

Scientists have long employed a set of systematic procedures for discovering *and* replicating findings, including persuasion findings. As a consequence, the study of persuasion no longer exists only as an ethereal art. It is now a science, a solid science, that can produce the same result time and again. What is more, whoever engages in the scientific process can duplicate the result. Brilliant, inspired individuals are no longer necessary to uncover the truth about persuasion. The power of discovery does not reside inside the minds of a few persuasive geniuses anymore but inside the scientific process. Therefore, knowledge about persuasion cannot be eliminated by eliminating the people who possess it—because somebody else can come along, use the same scientific procedures, and get the knowledge back again. So, anyone interested in becoming expert in the ways of persuasion is safe from threatened power holders, who should now be more interested in acquiring the information than abolishing it.

But, students of persuasion have a right to feel more than just relieved. They are entitled to feel encouraged by the fact that similar procedures can produce the same persuasion results over and over. If such replicability is indeed the case, it means that persuasion is governed by natural laws. The upshot offers a distinct advantage to anyone wishing to employ persuasion effectively. If persuasion is lawful, it is learnable. Whether born with an inspired talent for influence or not, whether preternaturally insightful about the process or not, whether a gifted artisan of the language or not, a person can learn how to be more influential. By applying a small set of principles that govern the persuasion process, one should be better able to move others in the direction of desired concessions, consensus, and compliance.

Six Universals of Persuasive Influence

For the past 30 years, I have been a fascinated participant in the search for a set of universal principles of persuasive influence, concentrating primarily on the major factors that bring about a specific form of behavior change—compliance with a request (Cialdini, 2001). What are the features of a request that my colleagues and I have found reliably spur a "yes" in response? Six central human tendencies appear to be key to successful influence of this sort: reciprocation, consistency, social validation, liking, authority, and scarcity.

Reciprocation

When the American Disabled Veterans' Organization sends out requests for contributions to potential donors in the United States, its appeal is productive about 18 percent of the time. But when the mailing includes an unsolicited gift (personalized address labels), the success rate jumps to 35 percent (Smolowe, 1990). Why? What is it about those gummed bits of paper, which no one requested and few desired, that could nearly double the effectiveness of the request? To understand, one must recognize the reach and power of an essential rule of human conduct: the code of reciprocity.

All societies subscribe to a norm that obligates individuals to repay in kind what they have received (Gouldner, 1960). When seen in this light, one can begin to appreciate why, upon receiving a packet of unwanted address labels from the veterans' organization, twice as many people would send a donation in return. It was not *what* they had received as a gift that was crucial. It was that they *had* received a gift.

Charitable organizations are far from alone in this approach. Food manufacturers offer free in-store samples, exterminators offer free in-home examinations, health clubs offer free workouts, and so on. The effect is not merely to give customers exposure to the product or service, it is also to indebt them. And the pull of the reciprocity rule extends beyond consumer decisions. Pharmaceutical companies spend millions of dollars per year to support medical researchers and to provide gifts to individual physicians. Evidence indicates that, as a result, researchers' findings and physicians' recommendations become drastically more favorable to these companies' interests. For instance, a 1998 study in the *New England Journal of Medicine* found that 37 percent of researchers who published conclusions critical of the safety of calcium-channel-blocking drugs had received prior drug company support; but every one of the researchers whose conclusions were favorable to the drugs' safety had received prior support in the form of free trips, research funding, or employment (Stelfox et al., 1998).

The rule for reciprocation does not just cover gifts and favors; it also applies to reciprocal concessions, that is, concessions that people make to one another. For instance, if you were to reject my large request and I then were to make a concession by retreating to a smaller request, you would likely reciprocate with a concession of your own—perhaps by agreeing to my smaller request. If you do not believe me, consider the results of an experiment my colleagues and I conducted. We stopped a random sample of passersby on public walkways and asked if they would be willing to volunteer to chaperone a group of inmates from the local juvenile detention center on a day trip to the zoo. As you can imagine, very few complied. But, for another random sample of passersby, we began with an even larger request—to serve as an unpaid counselor at the center for 3 h per week for the next 2 years! Not one of our second sample agreed to this extreme request. But, at that point, we offered them a concession, saying "Oh, if you can't do that, would you chaperone a group of juvenile detention center inmates on a day trip to the zoo?" That concession worked wonders, stimulating return concessions and nearly tripling compliance with the zoo trip request from 17 percent to 50 percent (Cialdini et al., 1975).

Consistency

Not long ago, Gordon Sinclair, the owner of a well-known Chicago restaurant, was struggling with a problem that afflicts all restaurateurs these days. Patrons frequently reserve a table but, without forewarning, fail to appear as scheduled. Mr. Sinclair solved the problem by asking his receptionist to change two words of what she said to callers requesting reservations—a change that dropped his "no-show" rate from 30 percent to 10 percent immediately. The two words were effective because they drew on the force of another potent human motivation: the desire to be (and to appear) consistent.

Most people prefer to be consistent with what they have previously done or said. For this reason, if I can get you to go on record, to make a public commitment, I will have greatly increased the chance that you will behave congruently with that commitment in the future. For example, Israeli researcher Joseph Schwartzwald and his coworkers were able to nearly double monetary contributions for the handicapped in certain neighborhoods by approaching residents 2 weeks before the actual request and getting them to sign a petition supporting the handicapped (Schwartzwald et al., 1983).

So, what were the two words that harnessed the tendency toward public consistency among Mr. Sinclair's restaurant patrons and pressed them to act in his interests? The receptionist modified her request from "Please call if you have to change your plans" to "*Will you* please call if you have to change your plans?" At that point, she paused politely … and waited for a response. To my mind, the wait was pivotal because it induced customers to fill the pause with a public commitment to comply with her request. And public commitments, even seemingly minor ones, direct future action.

Social Validation

One wintry New York morning, a man stopped for 60 s on a busy sidewalk and gazed skyward—at nothing in particular. He did so as part of an experiment by City University of New York social psychologists Stanley Milgram, Leonard Bickman, and Lawrence Berkowitz. It was designed to find out what effect this action would have on passersby. Most of them simply detoured or brushed by. Then, he did one thing differently that caused large numbers of pedestrians to halt, crowd together, and peer upward with him, still at nothing. What was it? I can offer two hints. First, he altered not one bit of what he did or said during that 60 s, staying stock-still and silent just as before. Second, the single change he made incorporated the phenomenon of "social validation."

One fundamental way that people decide what to do in a situation is to look to what others are doing or have done there. If *many* individuals like us have decided for a particular idea, we are more likely to follow, for we find the idea more correct, more valid, than would be the case without their lead. How did our New Yorker take advantage of the process of social validation to multiply his influence over passersby? He brought in four of his friends to stare skyward with him. When the initial set of upward-gazers increased from one to five, the percentage of New Yorkers who followed rose dramatically; and larger initial sets of friends generated even greater impact, nearly stopping traffic on the street within 1 min (Milgram et al., 1969). It appears that if numerous others seem to find merit in something—even something insubstantial—people assume that it must have merit, and they act accordingly.

As a result, requesters can foster our compliance by demonstrating (or merely implying) that others just like us have already complied. For example, in one study, a fundraiser who showed homeowners a list of neighbors who had donated to a local charity significantly increased the frequency of contributions; what is more, the longer the list, the greater was the effect (Reingen, 1982). It seems obvious, then, why marketers inform us that their product is the largest selling or fastest growing or why television commercials regularly depict crowds rushing to stores and hands depleting shelves of the advertised item.

Not so obvious, however, are the circumstances under which social validation can backfire. There is an understandable, but misguided, tendency of health educators to call attention to a problem by depicting it as regrettably frequent. Information campaigns stress that alcohol and drug use is intolerably high, that adolescent suicide rates are alarming, and that polluters are spoiling the environment. Although their claims are both true and well-intentioned, the creators of these campaigns have overlooked something basic about the compliance process: Within the statement

"Look at all the people who are doing this undesirable thing" lurks the powerful and undercutting message "Look at all the people who *are* doing it." Research shows that, as a consequence, many such programs boomerang, generating even more of the undesirable behavior. For instance, a suicide intervention program administered to New Jersey teenagers informed them of the alarming number of teenage suicides. Health researchers found that, as a consequence, participants became significantly more likely to see suicide as a potential solution to their problems (Shaffer et al., 1991). Much more effective are campaigns that honestly depict the unwanted activity as a damaging problem despite the fact that relatively *few* individuals perform it (Donaldson, 1995; Donaldson et al., 1995).

Liking

It is hardly surprising that people prefer to say yes to those they know and like. Consider, for example, the worldwide success of the Tupperware Corporation and its "home party" program. Through the in-home demonstration party, the company arranges for its customers to buy from and for a liked friend (the party hostess) rather than from an unknown salesperson. So favorable has been the effect on proceeds that, according to company literature, a Tupperware party begins somewhere in the world every 2.7 s.

But, of course, most commercial transactions do not take place in home parties among already-liked others. Under these much more typical circumstances, those who wish to invoke the power of liking must resort to another strategy: They must first get their influence targets to like them. How do they do it? The tactics that compliance practitioners employ cluster around certain factors that controlled research has shown to increase liking.

Physical attractiveness. Although it is generally acknowledged that good-looking individuals have an advantage in social interaction, most people sorely underestimate the size and reach of that advantage. For example, researchers found that voters in Canadian federal elections during the 1970s gave several times more votes to physically attractive candidates than to unattractive ones—while insisting that their choices would never be influenced by something as superficial as appearance (Efran & Patterson, 1974, 1976). Looks are influential in other domains as well. In a 1993 study conducted by Peter Reingen and Jerome Kernan, good-looking fundraisers for the American Heart Association generated nearly twice as many donations (42 percent versus 23 percent) as did other requesters.

Similarity. We humans like people who are similar to us. Thus, salespeople often search for (or fabricate) a similarity between themselves and their customers: "You're a skier? I love to ski!" Fundraisers do the same, with good results. For example, as part of one experiment, charity solicitors canvassed a college campus asking for contributions to a cause. When they added, "I'm a student, too" to their requests, donations more than doubled (Aune & Basil, 1994).

Compliments. Praise and other forms of positive estimation also stimulate liking. The simple information that one is appreciated can be a highly effective device for producing return liking and willing compliance. Indeed, praise may not have to be accurate to work. Research at the University of North Carolina found that compliments produced just as much liking for the flatterer when they were untrue as when they were genuine (Drachman et al., 1978). It is for such reasons that direct salespeople are trained in the use of praise.

Cooperation. Cooperation is another factor that has been shown to enhance positive feelings and behavior (Bettencourt et al., 1992). That is why compliance professionals often strive to be perceived as cooperating partners with a potential customer (Rafaeli & Sutton, 1991). Automobile sales managers frequently cast themselves as "villains" so that the salesperson can "do battle" on behalf of the prospective buyer. The cooperative, pulling-together kind of relationship that is consequently produced between the salesperson and customer naturally leads to a desirable form of liking that promotes sales.

Authority

Remember the man who used social validation to get large numbers of passersby to interrupt their progress and stare toward the sky with him? How might he use a different principle of influence or authority to accomplish the opposite? Rather than getting moving strangers to halt, how could he spur into motion stationary strangers waiting at a corner for a red light to change; and how could he do so without a single encouraging word or gesture? As discovered by a team of University of Texas researchers, the answer is simple: He could wear the right clothes. When he wore a suit and tie, which marked him as some kind of authority, 350 percent more pedestrians followed him across the street—against the light, against the traffic, and against the law—than when he was dressed casually (Lefkowitz et al., 1955).

Humans are not the only species to give sometimes single-minded deference to those in authority positions. In *The Social Contract* Robert Ardry (1970) reports on studies of food-taste acquisition in colonies of Japanese monkeys. In one troop, a taste for caramels was developed by introducing this new food into the diet of low-ranking members of the colony. A year and a half later, only 51 percent of the troop had acquired the taste, but still none of the leaders. Contrast this with what happened in a second troop where wheat was introduced first to the leader. Wheat-eating—to that point unknown to these animals—spread through the whole colony within 4 h.

Legitimate authorities are extremely influential in directing human conduct (Blass, 2000). Normally, it makes great sense to accept experts' guidance. Following their advice often helps facilitate rapid and correct choices. Therefore, people sometimes respond unthinkingly, deferring to an authority's judgment when it makes no sense at all: That Texas jaywalker, even in a suit and tie, was no more an authority on crossing the street than the rest of the pedestrians there. But when his clothing served as a symbol of authority, they followed.

It should come as no surprise that influence professionals frequently try to harness the power of authority by touting their experience, expertise, or scientific credentials: "In business since XXXX," "Four out of five doctors recommend the ingredients in XXXX," and so on. There is nothing wrong with such claims when they are real, for people usually want to know what true authorities think; it helps promote sound choices. The problem comes when phony claims are made. When people are not thinking hard, as is often the case when confronted by authority symbols, they can be easily steered in the wrong direction by ersatz experts—those who merely present the aura of legitimacy. For instance, several years ago in the United States, a highly successful ad campaign starred the actor Robert Young proclaiming the health benefits of decaffeinated coffee. Mr. Young appears to have been so effective in dispensing this medical opinion only because for many years he had played a physician (Marcus Welby, M.D.) on TV.

Scarcity

While a member of the faculty at Florida State University, psychologist Stephen West registered an odd occurrence after surveying students about the campus cafeteria cuisine. Ratings of the food rose significantly from the week before, even though there had been no change in the menu, food quality, or preparation. Instead, the shift resulted from an announcement that, because of a fire, cafeteria meals would not be available for several weeks (West, 1975).

This account highlights the impact of perceived scarcity on human judgment. A great deal of evidence shows that items and opportunities become more desirable as they become less available (Lynn, 1991). For this reason, marketers trumpet the unique benefits or the one-of-a-kind character of their offerings. It is also for this reason that they consistently engage in "limited time" promotions or put prospective consumers into competition with one another in "limited supply" sales programs.

Less widely recognized is that scarcity affects the value not only of commodities but of information as well. Information that is exclusive is more persuasive than information that is widely available. Take as evidence the dissertation data of a former student of mine, Amram Knishinsky—a man who owned a company that imported beef into the United States and sold it to supermarkets. To examine the effects of scarcity and exclusivity on compliance, he instructed his phone salespeople to call a randomly selected sample of customers and to make a standard request to purchase beef. He also instructed them to do the same with a second random sample of customers but to add that a shortage of Australian beef was anticipated, owing to certain weather conditions there. The added information that Australian beef was soon to be scarce more than doubled purchases. Finally, he instructed his salespeople to call a third sample of customers and to tell them about (a) the impending shortage of Australian beef and (b) the origin of this information—his company's *exclusive* sources in the Australian National Weather Service. These customers increased their orders by over 600 percent (Knishinsky, 1982). Why? Because they had received a scarcity one-two punch: Not only was the beef scarce, the information that the beef was scarce was itself scarce.

Defense

I think it is noteworthy that much of the data presented in this chapter has come from studies of the practices of the persuasion professionals. Who are the persuasion professionals and why should anyone find special insight in their approaches to the process of social influence? They are the individuals whose financial well-being depends on their ability to get others to say yes—marketers, advertisers, salespeople, fund-raisers, and the like. With this definition in place, one can begin to see why the regular practices of these professionals would lead one to the most powerful influences on the influence process—a law, not unlike natural selection, assures their emergence. Those practitioners who use unsuccessful tactics will soon go out of business, whereas those using procedures that work well will survive, flourish, and pass these successful strategies on—somewhat like adaptive genes—to succeeding generations (trainees). Thus, over time, the most effective principles of social influence will appear in the repertoires of long-standing persuasion professions. Those principles embody the six fundamental human tendencies examined in this article: reciprocation, consistency, social validation, liking, authority, and scarcity.

So, are people doomed to be the helpless victims of these principles? No. After all, in the vast majority of cases, the principles counsel correctly. Most of the time, it makes great sense to repay favors, behave consistently, follow the lead of similar others, favor the requests of likable others, heed legitimate authorities, and value scarce resources. Consequently, influence agents who use these principles honestly do consumers a favor. If an advertising agency, for instance, focused an ad campaign on the genuine weight of authoritative, scientific evidence favoring its client's headache product, all the right people would profit—the agency, the manufacturer, *and* the audience. Not so, however, if the agency, finding no particular scientific merit in the pain reliever, "smuggled" the authority principle into the situation through ads featuring actors wearing lab coats. The task of consumers, then, is to hold persuasion professionals accountable for the use of these six powerful motivators by purchasing their products and services, supporting their political proposals, and donating to their causes only when they have acted honestly in the process.

If we consumers make and enforce this vital distinction in our dealings with practitioners of the persuasive arts, we will rarely allow ourselves be tricked into assent. Instead, we will give ourselves a much better option: to be informed into yes. Moreover, as long as we apply the same distinction to our own influence attempts, we can legitimately avail ourselves of the same six principles in our campaigns for others' consent. In seeking to persuade by pointing to the presence of genuine expertise or growing social validation or pertinent commitments or real opportunities for cooperation and so on, we serve the interests of both parties and enhance the quality of the social fabric in the bargain. Helpless victims of the social influence process? Hardly.

References

Ardry, R. (1970). *The social contract*. New York: Antheneum.

Aune, R. K., & Basil, M. C. (1994). A relational obligations approach to the foot-in-the-mouth effect. *Journal of Applied Social Psychology, 24*, 546–556.

Bettencourt, B. A., Brewer, M. B., Croak, M. R., & Miller, N. (1992). Cooperation and the reduction of intergroup bias. *Journal of Experimental Social Psychology, 28*, 301–319.

Blass, T. (Ed.). (2000). *Obedience to authority: Current perspectives on the Milgram paradigm*. Mahwah, NJ: Erlbaum.

Cialdini, R. B. (2001). *Influence: Science and Practice* (4th ed.). Boston, MA: Allyn & Bacon.

Cialdini, R. B., Vincent, J. E., Lewis, S. K., Catalan, J., Wheeler, D., & Darby, B. L. (1975). Reciprocal concessions procedure for inducing compliance: The door-in-the-face technique. *Journal of Personality and Social Psychology, 31*, 206–215.

Donaldson, S. I. (1995). Peer influence on adolescent drug use: A perspective from the trenches of experimental evaluation research. *American Psychologist, 50*, 801–802.

Donaldson, S. I., Graham, J. W., Piccinin, A. M., & Hansen, W. B. (1995). Resistance-skills training and onset of alcohol use. *Health Psychology, 14*, 291–300.

Drachman, D., deCarufel, A., & Insko, C. (1978). The extra credit effect in interpersonal attraction. *Journal of Experimental Social Psychology, 14*, 458–465.

Efran, M. G., & Patterson, E. W. J. (1974). Voters vote beautiful: The effects of physical appearance on a national election. *Canadian Journal of Behavioral Science, 6*, 352–356.

Efran, M. G., & Patterson, E. W. J. (1976). *The politics of appearance*. Unpublished manuscript, Canada: University of Toronto.

Gouldner, A. W. (1960). The norm of reciprocity: A preliminary statement. *American Sociological Review, 25*, 161–178.

Knishinsky, A. (1982). *The effects of scarcity of material and exclusivity of information on industrial buyer perceived risk in provoking a purchase decision*. Unpublished doctoral dissertation, Tempe, AZ: Arizona State University.

Lefkowitz, M., Blake, R. R., & Mouton, J. S. (1955). Status factors in pedestrian violation of traffic signals. *Journal of Abnormal and Social Psychology, 51*, 704–706.

Lynn, M. (1991). Scarcity effects on value. *Psychology and Marketing, 8*, 43–57.

McGuire, W. J. (1985). Attitudes and attitude change. In G. Lindzey & E. Aronson (Eds.), *Handbook of social psychology* (3rd ed., Vol. 2, pp. 233–346). New York: Random House.

Milgram, S., Bickman, L., & Berkowitz, L. (1969). Note on the drawing power of crowds of different size. *Journal of Personality and Social Psychology, 13*, 79–82.

Rafaeli, A., & Sutton, R. I. (1991). Emotional contrast strategies as means of social influence. *Academy of Management Journal, 34*, 749–775.

Reingen, P. H. (1982). Test of a list procedure for inducing compliance with a request to donate money. *Journal of Applied Psychology, 67*, 110–118.

Reingen, P. H., & Kernan, J. B. (1993). Social perception and interpersonal influence: Some consequences of the physical attractiveness stereotype in a personal selling setting. *Journal of Consumer Psychology, 2*, 25–38.

Schwartzwald, J., Bizman, A., & Raz, M. (1983). The foot-in-the-door paradigm: Effects of second request size on donation probability and donor generosity. *Personality and Social Psychology Bulletin, 9*, 443–450.

Shaffer, D., Garland, A., Vieland. V., Underwood, M., & Busner, C. (1991). The impact of curriculum-based suicide prevention programs for teenagers. *Journal of the American Academy of Child and Adolescent Psychiatry, 30*, 588–596.

Smolowe, J. (1990, November 26). Contents require immediate attention. *Time*, p. 64.

Stelfox, H. T., Chua, G., O'Rourke, K, & Detsky, A. S. (1998). Conflict of interest in the debate over calcium-channel antagonists. *New England Journal of Medicine, 333*, 101–106.

West, S. G. (1975). Increasing the attractiveness of college cafeteria food. *Journal of Applied Psychology, 10*, 656–658.

Abstracts of the Contributions

Forms of Knowledge: Problems, Projects, Perspectives

Günter Abel

This chapter systematically describes different forms of knowledge and their roles at the interface of human cognition, communication, and cooperation (CCC triangulation). A distinction between a narrow and a broad sense of knowledge is made. The notion of forms of knowledge is explained as different ways of knowing. It is impossible to individuate contents of knowledge independently of signs and practices of articulation. There is no knowledge without signs. The author focuses on the relation between information and knowledge, following Kant in differentiating between "opinion," "belief," and "knowledge." In epistemological respects the chapter is an attempt to gain a foothold beyond the dichotomy of absolute knowledge and arbitrary relativism. The function of rules in knowledge acquisition and knowledge justification is shown to be particularly important. An object-oriented level is distinguished from a metatheoretical level, and it is shown in what sense second-order rules are embedded in and guaranteed by a world view. Hence, the power of world views, models, and systems of symbols in knowledge acquisition and knowledge dynamics is underlined. The dynamics of knowledge are brought into focus and internally correlated to symbols, time, situation, context, and creativity at the CCC interface. The role of nonpropositional, nonverbal, and implicit/tacit knowledge at the CCC interface is emphasized. The author outlines the internal relation between "knowing how" and "rationality," stressing that the rationality of knowing how is not algorithmic or calculus-guided; it is of a different type. An outline of a unified theory of knowledge and action is given.

Types of Sacred Space and European Responses to New Religious Movements

Eileen Barker

New Religious Movements (NRMs) have existed throughout history, but the visibility, extent, and variety of the movements in Europe, as elsewhere, has increased dramatically since World War II. This change is due partly to increased social and geographical mobility, partly to the exponential increase in the mass media, particularly the Internet, and partly to the general economic, political, and cultural globalizing tendencies of contemporary society. Whereas geopolitical boundaries have tended to circumscribe religious boundaries, the spread of alternative religions has resulted in unprecedented confusions and contestations over "what belongs where." The author considers locations of religious identity promoted from a variety of theological or ideological perspectives, including cosmic, global, national, local, biological, ethnic, lineage, cultural, individual, inner, and virtual space. She also compares reactions to religions that offer an alternative to the accepted orthodoxy within the countries of contemporary Europe and briefly outlines a variety of responses (both by individuals and by institutions such as governments, traditional religions, the media, and "cult-watching groups") to the question "What should be done about those minority religions?"

Turning Persuasion from an Art into a Science

Robert Cialdini

The winner of a clash of knowledge is often determined less by the features of the knowledge itself than by the way the knowledge is presented, with the winning side frequently being the one that makes the most persuasive presentation of its case. The most persuasive presentations are those that incorporate one or another of six universal principles of influence: (a) *Reciprocation*: People are more willing to comply with requests (e.g., for favors, services, information, or concessions) from those who have provided such things first. (b) *Commitment/Consistency*: People are more willing to be moved in a particular direction if they see it as consistent with an existing commitment. (c) *Authority*: People are more willing to follow the directions or recommendations of a communicator to whom they attribute relevant authority or expertise. (d) *Social Validation*: People are more willing to take a recommended action if they see evidence that many people, especially similar ones, are taking it. (e) *Scarcity*: People find objects and opportunities more attractive to the degree that they are scarce, rare, or dwindling in availability. (f) *Liking/Friendship*: People prefer to say yes to those they know and like.

Actors' and Analysts' Categories in the Social Analysis of Science

Harry Collins

Interpretative sociologists believe they must begin their analysis from the perspective of the actor. An analyst not rooted in the actors' world cannot explain it. Nevertheless, the analyst must eventually part with the actors, often challenging their perspective. There is no systematic work on how and when this shift should take place, but social studies of science increasingly indicate that it starts as a function of the political viewpoint of the analyst. Two classes of studies of science and technology are pointed out. In "traditional" studies of science and technology, the analyst shows that the science is much less clear cut than the way it is usually described for public consumption. In many more recent studies, especially those bearing on environmental issues, the science claiming that danger looms is argued by the analyst to be much *clearer* than some groups of scientists claim it to be. If social analysts of science are not forthright about how they make their choices as they switch from the actors' to the analysts' perspective, and if they are not careful to avoid self-serving choices, they will transform the subject. Instead of being an especially interesting and unique way of analyzing knowledge, their analyses will become the ideology of just another political pressure group.

The Theory of Cognitive Dissonance: State of the Science and Directions for Future Research

Peter Fischer, Dieter Frey, Claudia Peus, and Andreas Kastenmüller

The theory of cognitive dissonance has had profound impact on research in social psychology, Its influence goes beyond that field, however. The theory has, for example, also figured in the design of interventions to address a variety of societal problems. The chapter offers an overview of the empirical literature on dissonance theory, beginning with Festinger's (1957) classic definition. The authors cover the most important paradigms used in empirical dissonance research and summarize the most prominent empirical results. They explain the main features of the self-based revision of dissonance theory and introduce their own self-based modification of dissonance theory, including related data on ego-depletion and selective exposure. They conclude by outlining directions for future dissonance research, particularly on self-regulation and information-processing, and discuss the application of dissonance theory to societal problems.

Science and Religion in Popular Publishing in 19th-Century Britain

Aileen Fyfe

Modern belief in a controversy between science and religion has its origins in the activities of a relatively small number of intellectuals in the late 19th century. The author of this chapter aims to go beyond the intellectual circles, to consider how people in general thought about these issues. Religious practice was part of everyday life for a very large sector of the population, but there was nothing obviously equivalent for the sciences. The chapter focuses on popular publishing as one of the most significant ways in which nonspecialists could learn about the sciences. The author argues that, although secular presentations of the sciences were increasingly common in popular literature from the 1830s onwards, they did not represent opposition to religion per se, nor did Christian presentations disappear. Christian narratives of the sciences continued to appear (and to sell) long after professional science had been secularized. It is thus far from clear whether science did in fact replace theology as a system of knowledge for the majority of the population in the 19th century.

Cultural Boundaries: Settled and Unsettled

Thomas Gieryn

Whether or not the boundaries between cultural territories become occasions for contestation depends in part upon the architectural and geographical settings where they come together. At the Federal Building and Courthouse in Harrisburg, Pennsylvania, for example, centuries of disputation along the border between science and religion were reproduced yet again in 2005 when opposing parties disagreed on whether "intelligent design" should be included in the curriculum of public school science classes. By contrast, all seems calm at Stanford's Clark Center, a research facility (completed in 2003) where the potentially controversial boundary between science and politics is settled through the very design of the place. The Clark Center was built to materialize one particular set of political ambitions for science. Its spaces facilitate entrepreneurial, postdisciplinary, and rapidly reworked research. Yet most discourse surrounding the Clark Center avoids scrutiny of the boundary between science and politics, focusing instead on the building's aesthetic beauty and functional efficiency.

Reason, Faith, and Gnosis: Potentials and Problematics of a Typological Construct

Wouter Hanegraaff

This chapter contains an introduction to the academic study of Western esotericism, a new field of research that has been developing rapidly since the 1990s, and focuses on the role of "gnosis" in that context. Against an older approach associated chiefly with Gilles Quispel, the author argues that the triad of "reason—faith—gnosis" should not be used as a description of actual historical currents but that it may be useful as an analytical typology applicable to any kind of claimed knowledge. Whereas the first type of knowledge ("reason") is defined as both communicable and verifiable/falsifiable, and the second type ("faith") as communicable but not verifiable/falsifiable, gnosis is claimed to be a superior type knowledge that is neither communicable nor verifiable/falsifiable. The author argues that an adequate understanding of this third type requires cross-disciplinary methodologies that apply anthropological and psychological theories of "trance" or "altered states of consciousness" to the analysis of historical sources.

The Nexus Between Knowledge and Space

Peter Meusburger

The author debates some of the reasons why spatial disparities of knowledge evolve and why they are so persistent. The most prominent causes for spatial disparities of knowledge are the division of labor, the growth of complex social systems, the emergence of hierarchies, and the asymmetry of power relations in social systems. Before discussing relations between knowledge and space, the author inquires into concepts of space, place, spatiality, and spatial scales. He explains why many aspects of knowledge, education, and science cannot be perceived, described, and explained adequately if the spatial dimension is ignored. The proper consideration of spatial concepts and space–time has crucial effects upon the way theories and understandings are articulated and developed and upon the way the nexus of knowledge and space can be explicated. The author reviews the significance of spatial contexts for generating, legitimating, controlling, manipulating, and applying knowledge, especially scientific knowledge, and presents a brief report on the development and main research issues of geographies of education, knowledge, and science. The final part proposes a model for the spatial diffusion of various types of knowledge.

Science and the Limits of Knowledge

Mikael Stenmark

Some scientists have almost unlimited confidence in science and about what can be achieved in the name of science. For at least some of them, science even seems to be able to offer salvation. For others, science—that is, the natural sciences—at least sets the boundaries for what we humans can ever know about reality. This view is roughly that of scientism. This chapter gives an overview of different kinds of scientism. In particular, the focus is on the question about the limits of scientific knowledge. It is argued that scientism is a problematic position to take, one that in the end ought to be rejected. There are good reasons to believe that the world is bigger than the world of the natural sciences and that obtainable knowledge about this bigger world cannot be reduced to scientific knowledge.

When Faiths Collide: The Case of Fundamentalism

Roger Stump

Religious knowledge is rooted in two systems of meaning: a world view and an ethos. A religion's world view encompasses a cosmological understanding of reality, including conceptions of causation and agency and their relation to superhuman forces. A religion's ethos relates human existence to the reality defined in its world view, typically through basic norms, structures of daily life, and emotional patterns. The world view and ethos of a larger tradition, such as Christianity or Islam, repeatedly reflect processes of innovation tied to specific contexts as adherents transform them into local expressions. These processes take diverse forms, from unreflexive patterns of incremental change to explicit manifestations of schism and sectarianism. Fundamentalism represents an important form of religious innovation in the modern era, characterized by articulations of adherents' world view and ethos. Most importantly, fundamentalists are highly selective in defining the core elements of their world view, usually drawing on a literalist understanding of tradition but emphasizing some aspects of orthodoxy over others as a response to perceived threats to religious truth. The fundamentalist world view thus represents neither a complete rejection nor a precise recreation of earlier forms of orthodoxy. Fundamentalist world views particularly emphasize the perceived legitimacy of sources of truth and authority, producing systems of knowledge based on both faith and certainty. The worldly representation of such a system of knowledge, in turn, becomes central to the fundamentalist ethos, often provoking confrontations with others of the same religious traditions as well as with those outside it.

The Demarcation Problem of Knowledge and Faith: Questions and Answers from Theology

Michael Welker

This chapter critiques the use of the simple popular duality of "faith and knowledge." The religious and the academic realms that seem to represent it consist of truth-seeking communities and thus have strong structural similarities, although they are concerned with different subject matters. The typically modern achievement of a type of subjectivist faith, which has tried to fuse cognitive processes in the one realm with those in the other, in order to avoid any "clashes," has led to a systematic emptying of religious experience and communication. The author argues for a nondefensive understanding of the differences between religious and academic cognitive approaches in terms of their respective subject matters—amid deep similarities that should be acknowledged and appreciated.

The Klaus Tschira Foundation

Physicist Dr. h.c. Klaus Tschira established the *Klaus Tschira Foundation* in 1995 as a not-for-profit organization designed to support research in informatics, the natural sciences, and mathematics, as well as promotion of public understanding in these sciences. Klaus Tschira's commitment to this objective was honored in 1999 with the "Deutscher Stifterpreis" by the National Association of German Foundations. Klaus Tschira is a co-founder of the SAP AG in Walldorf, one of the world's leading companies in the software industry.

The Klaus Tschira Foundation (KTF) mainly provides support for research in applied informatics, the natural sciences, and mathematics, and supports educational projects for students at public and private universities and at schools. In all its activities, KTF tries to foster public understanding for the sciences, mathematics, and informatics. The resources provided are largely used to fund projects initiated by the Foundation itself. To this end, it commissions research from institutions such as the *EML Research*, founded by Klaus Tschira. The central objective of this research institute of applied informatics is to develop new information processing systems in which the technology involved does not represent an obstacle in the perception of the user. In addition, the KTF invites applications for project funding, provided that the projects in question are in line with the central concerns of the Foundation.

The home of the Foundation is the Villa Bosch in Heidelberg (Fig. 1), the former residence of Nobel Prize laureate for chemistry Carl Bosch (1874–1940). Carl Bosch, scientist, engineer and businessman, entered BASF in 1899 as a chemist and later became its CEO in 1919. In 1925 he was additionally appointed CEO of the then newly created IG Farbenindustrie AG and in 1935 Bosch became chairman of the supervisory board of this large chemical company. In 1937 Bosch was elected president of the Kaiser Wilhelm Gesellschaft (later Max-Planck-Gesellschaft), the premier scientific society in Germany. In his works, Bosch combined chemical and technological knowledge at its best. Between 1908 and 1913, together with Paul Alwin Mittasch, he surmounted numerous problems in the industrial synthesis of ammonia, based on the process discovered earlier by Fritz Haber (Karlsruhe, Nobel Prize for Chemistry in 1918). The Haber–Bosch Process, as it is known, quickly became and still is the most important process for the production of ammonia. Bosch's research also influenced high-pressure synthesis of other substances.

He was awarded the Nobel Prize for Chemistry in 1931, together with Friedrich Bergius.

In 1922, BASF erected a spacious country mansion and ancillary buildings in Heidelberg-Schlierbach for its CEO Carl Bosch. The villa is situated in a small park on the hillside above the river Neckar and within walking distance from the famous Heidelberg Castle. As a fine example of the style and culture of the 1920s it is considered to be one of the most beautiful buildings in Heidelberg and placed under cultural heritage protection. After the end of World War II the Villa Bosch served as domicile for high ranking military staff of the United States Army. After that, a local enterprise used the villa for several years as its headquarters. In 1967 the Süddeutsche Rundfunk, a broadcasting company, established its Studio Heidelberg here. Klaus Tschira bought the Villa Bosch as a future home for his planned foundations towards the end of 1994 and started to have the villa restored, renovated and modernised. Since mid 1997 the Villa Bosch has presented itself in new splendour, combining the historic ambience of the 1920s with the latest of infrastructure and technology and ready for new challenges. The former garage situated 300 m west of the villa now houses the Carl Bosch Museum Heidelberg, founded and managed by Gerda Tschia, which is dedicated to the memory of the Nobel laureate, his life and achievements.

This book is the result of a Symposium on "Clashes of Knowledge", which took place April 19–22, 2006, at the Villa Bosch (Fig.2).

For further information contact:

Klaus Tschira Foundation gGmbH
Villa Bosch
Schloss-Wolfsbrunnenweg 33
D-69118 Heidelberg, Germany
Tel.: (49) 6221/533-101
Fax: (49) 6221/533-199
beate.spiegel@ktf.villa-bosch.de

Public relations:
Renate Ries
Tel.: (49) 6221/533-214
Fax: (49) 6221/533-198
renate.ries@ktf.villa-bosch.de

www. ktf.villa-bosch.de

Fig. 1 The Villa Bosch (© Peter Meusburger, Heidelberg)

Fig. 2 Participants of the symposium "Clashes of Knowledge" at the Villa Bosch in Heidelberg, April 19–22, 2006 (© Thomas Bonn, Heidelberg)

Index

D

N

O

P

Q

R

Printed in the United Kingdom
by Lightning Source UK Ltd.
129299UK00002B/169-285/P

9 781402 055546